Air Attack Against Wildfires

Understanding U.S. Forest Service Requirements for Large Aircraft

Edward G. Keating, Andrew R. Morral, Carter C. Price,
Dulani Woods, Daniel M. Norton, Christina Panis,
Evan Saltzman, Ricardo Sanchez

Sponsored by the United States Forest Service

RAND HOMELAND SECURITY AND DEFENSE CENTER

This research was sponsored by the United States Forest Service and was conducted within the RAND Homeland Security and Defense Center, a joint center of the RAND National Security Research Division and RAND Infrastructure, Safety, and Environment.

Library of Congress Control Number: 2012944768

ISBN: 978-0-8330-7677-9

Published 2012 by the RAND Corporation
1776 Main Street, P.O. Box 2138, Santa Monica, CA 90407-2138
1200 South Hayes Street, Arlington, VA 22202-5050
4570 Fifth Avenue, Suite 600, Pittsburgh, PA 15213-2665
RAND URL: http://www.rand.org/
To order RAND documents or to obtain additional information, contact
Distribution Services: Telephone: (310) 451-7002;
Fax: (310) 451-6915; Email: order@rand.org

Preface

In the fall of 2009, the U.S. Forest Service, an agency of the U.S. Department of Agriculture, asked the RAND Corporation to undertake the study "Determination and Cost-Benefit Analysis of the Optimal Mix of Helicopters and Airtankers for the U.S. Forest Service." The Forest Service requested additional follow-up analyses in the fall of 2010. The objective of these research projects was to assist the Forest Service in determining the composition of a fleet of airtankers, scoopers, and helicopters that would minimize the total social costs of wildfires, including the cost of large fires and the cost of aircraft. This report summarizes the research approach and results and should be of interest to Forest Service officials and others who are concerned with ensuring that the nation's wildfire-fighting capabilities are maintained in an efficient and cost-effective way.

The RAND Homeland Security and Defense Center

This research was conducted in the RAND Homeland Security and Defense Center, which conducts analysis to prepare and protect communities and critical infrastructure from natural disasters and terrorism. Center projects examine a wide range of risk management problems, including coastal and border security, emergency preparedness and response, defense support to civil authorities, transportation security, domestic intelligence, technology acquisition, and related topics. Center clients include the Department of Homeland Security, the Department of Defense, the Department of Justice, and other organizations charged with security and disaster preparedness, response, and recovery. The Homeland Security and Defense Center is a joint center of the RAND National Security Research Division and RAND Infrastructure, Safety, and Environment.

Questions or comments about this report should be sent to the project leader, Edward G. Keating (Edward_Keating@rand.org). Information about the Homeland Security and Defense Center is available online (http://www.rand.org/multi/homeland-security-and-defense/). Inquiries about homeland security research projects should be sent to:

Andrew Morral, Director
Homeland Security and Defense Center
RAND Corporation
1200 South Hayes Street
Arlington, VA 22202-5050
703-413-1100, x5119
Andrew_Morral@rand.org

Contents

Figures

Tables

Summary

An aging fleet of contracted fixed-wing airtankers and two fatal crashes of these aircraft led the U.S. Forest Service, an agency of the U.S. Department of Agriculture, to investigate the cost of obtaining new airtankers. The Forest Service asked the RAND Corporation for assistance in determining the composition of a fleet of airtankers, scoopers, and helicopters that would minimize the total social costs of wildfires, including the cost of large fires and aircraft costs. RAND was not asked to consider whether the Forest Service should own or contract for its firefighting aircraft.

Background

Wildland fires are among nature's most terrifying and dangerous phenomena. At the same time, periodic wildfires are a natural part of ecosystem dynamics in much of the country. Policymakers therefore face difficult choices as to how, whether, and to what extent to fight wildland fires when they break out.

On-the-ground firefighters can create "fire lines" to contain a blaze. A fire line is a buffer of cleared or treated ground that resists a fire's further growth. When a fire is encircled by a fire line, it is said to be contained. Firefighting aircraft can abet efforts to build a fire line by dropping retardant, suppressant, or water on burning or potential fuel.

While the use of aircraft often garners media attention, there is a dearth of empirical evidence that aircraft are effective against already-large fires. There is much firmer evidence that aircraft can assist in the "initial attack," i.e., support on-the-ground firefighters in containing a potentially costly fire while it is still small.

The focus of this analysis is on large aircraft, a term we use to denote Type I helicopters (those that can lift 5,000 or more pounds), as well as 1,500- to 3,000-gallon airtankers and scoopers. We did not consider the use of military-operated aircraft (e.g., Air Force C-130 cargo aircraft) in firefighting.

The Cost of Large Fires

Large fires can be enormously costly to fight and can result in sizable damages. A successful initial attack saves the public these costs, or at least slows their accumulation. Estimating the cost of large fires requires tabulating the available data on federal suppression expenditures, the state and local suppression costs of fires to which the Forest Service has responded, federal post-fire rehabilitation costs, insured losses, fatalities, and future suppression costs.

We estimated that a large fire has an average social cost of between $2.1 million and $4.5 million. We present this range of values to account for uncertainties in key parameters, including suppression costs and the magnitude of insured losses.

Our baseline estimate is that a successful initial attack that prevents a large fire saves $3.3 million, the average of $2.1 million and $4.5 million.

The Cost of Large Aircraft

Our analysis also involved tabulating the cost of prospective aircraft. We analyzed five candidate aircraft in three categories: 1,500-gallon and 3,000-gallon airtankers, a 1,600-gallon scooper, and 1,200-gallon and 2,700-gallon helicopters.

We estimated aircraft life-cycle costs using information drawn from multiple publicly available sources. Note that they are not "source selection–quality" cost estimates. They are best estimates based on available information, but they have not been verified with the contractors.

We estimate an annualized life-cycle cost per aircraft in the range of $3 million to $8 million, not including retardant costs. Not surprisingly, larger airtankers have higher annualized costs. Retardant costs further increase the cost of airtankers relative to helicopters and scoopers, which do not generally drop retardant.

The RAND National Model

We used two separate but complementary models to estimate the social cost–minimizing portfolio of initial attack aircraft. The RAND National Model is an optimization model that views aircraft allocation as a national problem. It identifies options for relocating aircraft at the national level to stop as many small fires as possible from becoming large and costly.

The National Model's fire simulation is based on data on wildfires in the United States in calendar years 1999–2008. Using the model, an analyst can run different prospective portfolios of aircraft against these historical fires to assess how outcomes might have differed with more or fewer available aircraft. The National Model assumes that a fixed, baseline level of local fire-line production capability is available against any given fire and that aircraft supplement these local resources.

We identified three types of small fires. Category A fires are those that will be contained by local resources, even without aviation support. A Category B fire will become large if only baseline local fire-line production resources are used, but large aircraft can augment those resources to achieve containment. A Category C fire will become large irrespective of large aircraft usage. If aircraft dispatchers had perfect information about every small fire (a condition we refer to as *dispatch prescience*), aircraft would be sent against Category B fires only. However, in the interest of realism, we assumed that aircraft, if they are available, are sent to all Category B fires, as well to all Category C fires and to "close-call" Category A fires (i.e., those that have rates of growth close to that of a Category B fire).

We found that at least two-thirds of historical fires have been within ten miles of a scooper-accessible body of water, and about 80 percent have been within five miles of a helicopter-accessible body of water. These water-proximate fires are the fires against which helicopters and scoopers would be most valuable. Our base assumption, in accordance with findings from Fire Program Solutions (2005), is that water is half as effective as retardant on a per-gallon basis.

Our baseline National Model simulation suggests an optimal initial attack fleet of five 3,000-gallon airtankers and 43 1,600-gallon scoopers.

These results are sensitive to assumptions about the relative effectiveness of water versus retardant. However, we found that it would take a considerable degradation in the efficacy of water relative to retardant for an airtanker-centric, rather than a scooper-centric, portfolio to be preferred in the National Model simulation.

The RAND Local Resources Model

A major concern with regard to the National Model is that it assumes that local firefighting resources are the same across the country. To loosen this assumption, we additionally developed the RAND Local Resources Model. This model uses data on local firefighting resources to characterize the impact of a given mix of aircraft against specific small fires. The Local Resources Model also allows the social costs of large fires to vary by location.

To incorporate information on local firefighting resources, the Local Resources Model relies on data and model results from the Fire Program Analysis (FPA) system, a Forest Service system designed to assist decisionmakers with resource allocation choices. Although FPA's traditional focus has been at the local U.S. Forest Service Fire Planning Unit level, the Local Resources Model uses the system to estimate the outcomes of specific fires in relation to the availability of large aircraft. In the Local Resources Model, it is FPA's algorithms that determine whether an air attack with retardant or water changes a fire's outcome (specifically, from large to small). To these FPA algorithms, the Local

Resources Model adds accounting and optimization algorithms that track the availability of a fixed fleet of aircraft of specified types and known locations. The Local Resources Model dispatches aircraft to fires simulated by the FPA model, calculates aircraft cycle times and fuel endurance, and, in the case of scoopers and helicopters, determines the nearest available water sources. The Local Resources Model can be used to investigate the number of large fires—and the resulting social costs—associated with alternative airtanker fleet sizes and mixes.

Our baseline assessment using the Local Resources Model suggests an optimal initial attack fleet composed of one 3,000-gallon airtanker, two 2,700-gallon helicopters, and 15 1,600-gallon scoopers.

We performed sensitivity analyses for the factors for which limited data were available, including the efficiency with which the Forest Service can pre-position its aircraft and the prescience with which it dispatches aircraft only to the fires that are most likely to be contained (i.e., the small fires that the aircraft can prevent from becoming large fires). If the Forest Service has sufficient insight into where fires will next occur, has the freedom to move its resources to any airport that can optimize an attack, and is precise in sending aircraft only to those fires that require aircraft for containment, the total size of the required fleet would be substantially smaller than if the Forest Service had poorer intelligence on future fires, less flexibility in pre-positioning aircraft, or less insight into which fires were most appropriate for aircraft to fight.

It is important to note that our analysis is subject to the limitations of the Forest Service's FPA model, on which the Local Resources Model is built. Specifically, FPA makes important assumptions about the tactical equivalence of water versus retardant, the tactical role of aircraft in initial response (for instance, that aircraft cannot slow or suppress fires in advance of the arrival of ground resources), and the tactical equivalence of attacking the burning edge of a fire with water or retardant ("direct attack") and building a fire-control line away from the burning edge ("indirect attack"). If these assumptions are invalid or only partially correct, the utility of the results of the Local Resources Model will be correspondingly limited.

Both the National Model and the Local Resources Model analyze aircraft to be used in an initial attack. Unfortunately, little is known about the value of aircraft against already-large fires. However, if assumptions could be made about the daily value of aircraft against an already-large fire, it would be possible to estimate how many additional aircraft should be acquired, beyond those acquired for use in the initial attack phase.

Concluding Remarks

Both models have important limitations due to the unavailability of key data or established science and the need to make sometimes-important assumptions. Given their different underlying assumptions, it is not surprising that the National Model and the Local Resources Model produce different estimates of optimal initial attack aircraft portfolios, as shown in Table S.1. The shaded cells in Table S.1 represent the models' respective base-case estimates.

Rather than trying to adjudicate which (if any) of these findings is "best" or "right," we draw broader insights from the models' different results. These insights may be helpful to the Forest Service's leadership as it considers the fleet mix that is most likely to optimize taxpayer returns on investment.

In each case, scoopers are the central component of the optimal solution. Two factors drive this finding. First, scoopers are considerably less expensive to own and operate than larger helicopters and fixed-wing airtankers. Second, when fires are proximate to water sources, scoopers can drop far more water on a fire than a retardant-bearing aircraft can drop retardant. Because most human settlement is near water, scoopers can be highly effective against many of the most costly fires.

Retardant-bearing airtankers are also valuable, but primarily in the niche role of fighting the minority of fires that are not water-proximate.

In developing both the National Model and the Local Resources Model, we confronted the issue of dispatch prescience and its importance in determining optimal initial attack aircraft portfolios. Greater dispatch prescience can sharply reduce required initial attack air-

Table S.1
Estimates of Optimal Initial Attack Fleets in the National and Local
Resources Models

Case	RAND National Model	RAND Local Resources Model
Water-retardant efficacy parity	2 airtankers, 40 scoopers	1 airtanker, 15 scoopers, 2 helicopters
Water half as effective as retardant	5 airtankers, 43 scoopers	X
Water one-quarter as effective as retardant	9 airtankers, 43 scoopers	X
$2.1 million per average large fire	4 airtankers, 36 scoopers	2 airtankers, 14 scoopers, 2 helicopters
$4.5 million per average large fire	6 airtankers, 55 scoopers	1 airtanker, 14 scoopers, 4 helicopters
Geographical constraint	8 airtankers, 48 scoopers	4 airtankers, 25 scoopers, 7 helicopters

NOTE: The National Model geographical constraint case restricts aircraft to operating in a single Forest Service Geographic Area Coordinating Center in a given month. The Local Resources Model geographical constraint case has each aircraft assigned to a given base for at least 20 days. The Local Resources Model does not allow varying water efficacy. The shaded cells represent each model's base-case estimate.

craft portfolios. The importance of prescience suggests the possibility of reducing the required number of aircraft or reducing the number of large, costly fires by improving aircraft assignment and dispatch algorithms.

Given the frequency with which large airtankers are used against already-large fires, we were surprised by the dearth of statistical evidence documenting their value in this role. Better information about the costs and benefits of air assault in large fire operations would help clarify the optimal mix of aircraft required for wildland firefighting operations.

There are several possible extensions of this work, including

- allowing water-efficacy parameter flexibility in FPA and, hence, in the Local Resources Model
- bringing local resource data from FPA into the National Model

- analyzing how Forest Service aircraft have been used, e.g., their patterns of relocation (where and how frequently), the amount of time they spend fighting small versus already-large fires
- assessing, perhaps experimentally, how often aircraft truly change outcomes between small and large fires
- calibrating the frequency and efficacy of direct versus indirect attack in today's airtanker fleet.

Any of these extensions would abet further efforts to understand the Forest Service's requirements for large aircraft.

Acknowledgments

We would like to extend our appreciation for the research sponsorship of Larry Brosnan and Paul Linse of the U.S. Forest Service. Each provided numerous comments and suggestions that assisted in this research.

The RAND study team benefited from a December 4, 2009, visit to Boise, Idaho, and meetings with Forest Service personnel there, including Joe Frost, Laura Hill, and Tom Zimmerman. We met with Paul Linse, Scott Fisher, Tory Henderson, and Colleen Hightower in Boise on April 21, 2010. We also benefited from a May 25, 2010, visit to Missoula, Montana, and meetings with Forest Service personnel there, including Dave Calkin, Kathy Elzig, Krista Gebert, Bob Roth, Keith Stockmann, Kim Thomas, Matthew Thompson, Diane Trethewey, and Shirley Zylstra.

The authors are particularly grateful for insightful exchanges with Patrick Basch, Dan Crittenden, Margaret Doherty, Jeremy Fried, Art Hinaman, Patti Hirami, Neal Hitchcock, Richard Kvale, Charlie Leonard, John Nelson, Marc Rounsaville, Aaron Schoolcraft, Brent Timothy, and Khon Viengkham of the Forest Service. In addition, Joe Frost, Cal Gale, Carmi Gazit, and Doris Lippert provided extensive assistance with the FPA system, and Brad Harwood provided data for the models. Scott Fisher and Colleen Hightower also assisted with contracting issues.

Dave Calkin, Jeremy Fried, Paul Linse, and Matthew Thompson provided helpful comments on a draft of this report. We also received constructive reviews of an earlier version from RAND colleagues John

Birkler, Lloyd Dixon, and Richard Hillestad and from Jonathan Yoder of Washington State University.

We are grateful for insights provided by Thomas Y. Eversole, executive director of the American Helicopter Services and Aerial Firefighting Association, at a January 26, 2010, meeting, and Joel Kerley of the U.S. Department of the Interior, with whom we met on September 22, 2010. Todd Peterson, vice president of Columbia Helicopters, also provided helpful guidance for our research

We received helpful feedback and input on relevant Australian experiences from Jim Gould and Matt Plucinski of Australia's Commonwealth Scientific and Industrial Research Organisation.

Martin Alexander of Natural Resources Canada provided copies of several of his research papers that we had trouble obtaining.

John Gonsalves of Bombardier connected us with numerous well-informed users of CL-215 and CL-415 scooper aircraft. We also appreciate the insights offered by John Gould from the Bureau of Land Management in Alaska, Ken Giesbrecht from the Province of Manitoba, Sheldon Mack from the State of Minnesota, Eric Earle from the Province of Newfoundland, Jerry Kroetsch from the Province of Ontario, Brian Fennessy from San Diego County, and Steve Roberts from the Province of Saskatchewan. We especially enjoyed a February 8, 2011, meeting with Los Angeles County firefighters James Crawford, Anthony Marrone, and Steve Martin.

RAND Infrastructure, Safety, and Environment vice president and director Debra Knopman provided comments on earlier versions of this document and was consistently supportive of this project. Tom LaTourrette and Di Valentine coordinated the document review process, and Henry Willis provided additional helpful comments. Neil DeWeese assisted with project management, Yvonne Montoya and Joye Hunter assisted in proposal preparation, and Nick Byone assisted with contract management. Lauren Skrabala edited this document and prepared its accompanying research brief.

RAND colleagues Mark Arena and John Schank provided support as we formulated our research proposal and served as sounding boards as the research moved forward. Winfield Boerckel and Laura Selway facilitated legislative interactions. Carol Fan, Elvira Loredo, and

Ronald McGarvey provided operations research advice. John Halliday suggested that we create the graphical representation of gallons of water dropped relative to the aircraft's proximity to a body of water, which became Figure 5.3. Jeffrey Sullivan assisted with the modeling. Aaron Kofner, Judith Mele, Lisa Miyashiro, Adrian Overton, and Chuck Stelzner provided geographic information system assistance; Miyashiro created Figure 5.8. Robert Leonard, Gary Massey, and Fred Timson helped with our cost analyses. Susan Marquis and Lisa Sodders noted relevant literature, and Christopher Mouton provided insights on helicopter performance as a function of elevation. We also benefited from Greg Ridgeway's statistical assistance, Jerry Sollinger's communication assistance, and Robert Weissler's programming assistance.

Tomiko Envela and Brooke Hyatt fielded our requests for library resources and literature. Suzanne Benedict, Alex Chinh, and Jerome Hawkins provided administrative assistance. Lisa Bernard, Todd Duft, Stacie McKee, and Benson Wong helped prepare this report for publication.

Of course, the research findings, interpretations, and any remaining errors are solely the authors' responsibility.

Abbreviations

DOI	U.S. Department of the Interior
FAA	Federal Aviation Administration
FIG	Fire Ignition Generator
FPA	Fire Program Analysis
GACC	Geographic Area Coordination Center
IRS	Initial Response Simulator
NIFC	National Interagency Fire Center
O&S	operating and support
OMB	Office of Management and Budget
SCI	Stratified Cost Index
WUI	wildland-urban interface

CHAPTER ONE

Introduction

The U.S. Forest Service, an agency of the U.S. Department of Agriculture, has an aging fleet of contracted fixed-wing airtankers to assist in fighting wildfires. Tragically, there were two fatal crashes of Forest Service–contracted airtankers in 2002. On June 17, 2002, a C-130A experienced an in-flight breakup initiated by the separation of the right wing, followed by the separation of the left wing, while executing a fire retardant drop over a forest fire near Walker, California. All three flight crewmembers were killed, and the airplane was destroyed. On July 18, 2002, a Forest Service–contracted P4Y aircraft experienced an in-flight separation of the left wing while maneuvering to deliver fire retardant over a forest fire near Estes Park, Colorado. Both flight crewmembers were killed, and the airplane was destroyed.[1]

After these crashes, the remainder of the Forest Service's contracted airtanker fleet was grounded. Ultimately, fewer than half of the fleet of 44 2,000- to 3,000-gallon airtankers returned to service. These remaining 19 contracted airtankers have a limited remaining service life, and the Forest Service plans to replace them over the next few years.

The Office of Management and Budget (OMB) rejected an earlier Forest Service proposal to replace the aircraft on the grounds it lacked both an acquisition plan and a sufficient cost-benefit analysis justifying the need for airtankers. The U.S. Department of Agriculture's Office

---

[1] A more detailed account of the crashes can be found in National Transportation Safety Board (2004).

of the Inspector General (2009) stated that the subsequent cost-benefit analysis required a more persuasive justification for new aircraft.

In response, the Forest Service asked the RAND Corporation for assistance in determining the composition of a mix of airtankers, scoopers,[2] and helicopters that would minimize the total costs of wildfires, including the cost of large fires and the cost of aircraft.

One might view our study as a total cost minimization exercise. The goal is to choose the portfolio of firefighting aircraft, a, that minimizes the total cost. The total cost consists of the sum of fire-related costs, C_F, and aircraft-related costs, C_A. The cost of fires, C_F, is a function of the wildfires that occur, f. But f itself is a function of a variety of independent variables (e.g., weather, pre-suppression tactics), including the number of aircraft: Having more aircraft reduces the number and costs of wildfires. Of course, having more aircraft also increases aircraft costs. The overall objective is to choose the number of aircraft to minimize the sum of costs of fires and aircraft. These values trade off on one another, i.e., a large portfolio of aircraft would reduce fire costs but would imply large aircraft costs.

We can express this exercise in mathematical notation, where f is a function of a variety of variables, including a, where having more aircraft reduces the number and cost of wildfires but increases the cost of the fleet. The overall objective function is to choose a to minimize $C_F(f(a)) + C_A(a)$, recognizing that a large portfolio, a, reduces fire costs but increases aircraft costs.

Other researchers have examined the value of aviation in fighting wildfires. For example, Countryman (1969) presented a case study of airtanker efficacy in fighting a 1967 fire in the Los Padres National Forest in Southern California. He found that airtankers increased suppression costs, but this was justified by reduced acres burnt and, hence, reduced damages. Martell et al. (1984) evaluated initial attack resources in forest fires in Ontario, Canada, and Loane and Gould (1986) undertook a detailed cost-benefit study for the Australian state of Victoria on the aerial suppression of bushfires.

---

[2]   Scoopers, as the name implies, scoop water out of lakes, rivers, or the ocean and then drop it on fires. Scoopers, unlike rotary-wing helicopters, are fixed-wing aircraft.

The Forest Service found that the primary need for both Type I (large) and Type II (medium-sized) helicopters is in supporting large fire suppression operations (U.S. Department of Agriculture, Forest Service, 1992).[3] In 1995, the Forest Service and the U.S. Department of the Interior (DOI) recommended a national fleet size of 41 large airtankers (U.S. Department of Agriculture, Forest Service, and U.S. Department of the Interior, 1995). A follow-up study by the two agencies (1996) recommended the procurement of excess military aircraft, suggesting a fleet composed of 20 P-3A aircraft, ten C-130B aircraft, and 11 C-130E aircraft. Fire Program Solutions (2005) found that airtanker platforms of 3,000–5,000 gallons were significantly more cost-effective than smaller-capacity platforms. Its study suggested that airtankers are more efficient than helicopters in building the fire line in an initial attack on small fires but that helicopters are preferred for large fire support.

There has been considerable modeling, research, and evidence collection related to the value of aircraft in the initial attack phase.[4] *Initial attack* refers to fighting fires while they are small to prevent them from becoming large and much more costly. There is far less evidence of the benefits of aircraft against already-large fires. Therefore, we approached the task of determining the optimal mix of large aircraft in two phases. In the first phase, we modeled the effects of alternative fleet mixes in an initial attack. In the second phase, the results of which are presented at the end of Chapter Six, we considered the benefits that must be assumed to accrue from a large fire attack to warrant the acquisition of additional aircraft beyond those selected for the initial attack.

This report presents the results of two models that we call the RAND National Model and the RAND Local Resources Model. The National Model is an optimization model that views aircraft allocation as a national problem, with aircraft allocated at the national level

---

[3] Personal communication with Paul Linse, U.S. Forest Service, July 17, 2012, defined a Type I helicopter as one that can lift 5,000 or more pounds, a Type II helicopter as one that can lift 2,500–4,999 pounds, and a Type III helicopter as one that can lift 1,200–2,499 pounds.

[4] See, for example, Bradstock, Sanders, and Tegart (1987); U.S. Department of Agriculture, Forest Service (1992); U.S. Department of Agriculture, Forest Service, and U.S. Department of the Interior (1995); Fried and Fried (1996); and McCarthy (2003).

to stop as many small fires as possible from becoming large and costly. The National Model trades off the cost of aircraft (having more aircraft increases costs) against the costs of large fires (having more aircraft results in fewer large fires).

Our greatest concern about the National Model is that it does not account for differential local firefighting resources. Some parts of the country (e.g., Los Angeles County) have considerable local firefighting resources, but other areas (e.g., eastern Nevada) have relatively few. The marginal impact of a Forest Service firefighting aircraft would therefore be different in different areas. Of course, there are good reasons for greater firefighting resources in Los Angeles County: The area at risk is much more densely populated and has high-value buildings and infrastructure. Ideally, an aircraft optimization model would account for differences in both the local firefighting resources available and the value at risk.

We developed the Local Resources Model to address the lack of local resource consideration in the National Model. The Local Resources Model uses data on local firefighting resources in the Forest Service's Fire Program Analysis (FPA) system. FPA simulates fires and the resulting initial attack outcomes given local firefighting resources with or without Forest Service large firefighting aircraft. The Local Resources Model uses the FPA simulation results to determine optimal initial attack aircraft fleet sizes and locations, trading off the costs of large aircraft against the costs of large fires.

The Local Resources Model is not without concerns. Most centrally, it is dependent on the validity of the FPA simulations. Forest Service personnel raised concerns about latent assumptions in the system. For instance, FPA attributes as much efficacy to a gallon of water dropped from a scooper as to a gallon of retardant dropped from an airtanker. This assumption is contrary to a traditional assumption that retardant is twice as effective as water on a per-gallon basis,[5] but

---

[5] Fire Program Solutions (2005, p. 16) highlights the traditional two-to-one retardant-to-water efficacy assumption. The assumption is supported by performance results in a standard burn test dated January 16, 2008, provided to RAND in personal communication from Tory Henderson of the Forest Service on April 26, 2010.

we were not able to modify FPA's inherent parameters. Instead, the Local Resources Model sits astride FPA, with FPA being, de facto, a "black box" from an analysis perspective. We were, however, able to use the more flexible National Model to assess the impact of different assumptions of retardant-to-water efficacy.

Not surprisingly, the National Model and the Local Resources Model provide different point estimates as to the Forest Service's optimal initial attack aviation fleet. Rather than assessing which model is "better" or "right," we think it is more constructive to consider some of their broader lessons and consistencies:

- Both models suggest that scoopers have the central role in initial attack, even though water dropped from a scooper is half as effective as retardant dropped from a fixed-wing airtanker in the National Model (appropriately, in our view). The key virtue of scoopers is that they can drop far more water per hour on most fires than airtankers can drop retardant. Our analysis of geographic information system data shows that most high-risk fires occur near water sources, precisely because most human settlement is near water.
- Access to fixed-wing airtankers is also valuable in the minority of fires that are not proximate to water sources. Furthermore, some airtanker availability is a useful hedge against a scenario in which scoopers may lack permission to draw off a proximate water source.
- There is a trade-off between the number of aircraft needed (of any type) and the prescience with which those aircraft are used. If the Forest Service used firefighting aircraft only when the aircraft would be most effective in preventing a large, costly fire, only a small fleet would be needed. But dispatchers lack such perfect information. We cannot expect aircraft dispatchers to know exactly which small fires will benefit from aircraft and which will not. As aircraft dispatch becomes less prescient, more aircraft are needed. This phenomenon suggests an opportunity for better strategic decisions about aircraft locations (e.g., where to preposition firefighting aircraft) and better tactical decisions about

aircraft usage (e.g., fires to which aircraft should be sent) to reduce required investments in firefighting aircraft.

Neither the National Model nor the Local Resources Model considers aircraft usage against already-large fires. As mentioned earlier, there is a paucity of evidence as to the benefits of aircraft against already-large fires. If, however, one is willing to make an assumption of the daily value of such aircraft against large fires, one can then calculate an appropriate augmentation to the initial attack fleets suggested by the National Model and the Local Resources Model.

The remainder of this report is structured as follows: Chapter Two provides background information on wildfires and firefighting. Chapter Three discusses the social costs of large fires. Chapter Four discusses the estimated costs of acquiring, operating, and maintaining a national fleet of large aircraft. Chapter Five and Chapter Six then present findings from our minimization explorations: Chapter Five presents results from the National Model and its estimation of the optimal initial attack fleet, and Chapter Six presents the results from the Local Resources Model. We also discuss how the Local Resources Model's findings compare with those from the National Model. Chapter Seven presents concluding remarks. Two appendixes provide additional detail on the analyses underlying the cost estimates in Chapter Three and future trends in the use of aviation assets to fight wildfires, respectively.

CHAPTER TWO

Background

Wildland fires are among nature's most terrifying and dangerous events. At the same time, periodic wildland fires are a natural and beneficial part of the ecosystem in much of the country. Many species of flora and fauna rely on episodic wildfire as part of their regenerative cycle (see, e.g., Agee, 1989; Servis and Boucher, 1999). Nevertheless, because fires threaten property, public health, and forest resources, there is a strong interest in preventing and suppressing them. Policymakers therefore face difficult choices about how, whether, and to what extent to fight wildland fires that break out.

There are several ways to contain the damage caused by wildland fires. The most passive approach is to do nothing beyond evacuating people from affected areas. Fires will eventually be extinguished when they run out of fuel, when the weather changes, or when they encounter geographic barriers, such as lakes or rivers.

More proactively, on-the-ground firefighters can create a "fire-control line" to contain a fire. A fire line, like a lake or river, serves as a barrier to a fire's further advancement. When a fire is encircled by fire line, it is said to be contained. Although it may continue to burn, a contained fire has a reduced risk of consuming areas outside the fire line (unless, for instance, embers or flames jump beyond the line, igniting areas outside the contained area). Dropping retardant at strategic locations in the path of the burning edge of a fire would be an "indirect attack."

The alternative to an indirect attack is a "direct attack," in which firefighters build a fire line at the burning edge of a fire. The National

Wildfire Coordinating Group (2004, p. 92) notes that a direct attack approach is used when the fire perimeter is burning at low intensity and fuels are light, allowing safe operation at the fire's edge. The indirect attack approach offers firefighters more protection from smoke and heat.

Traditionally, a fire line is created by firefighters using tools, such as axes or bulldozers, to clear brush and trees or fire trucks and water hoses to moisten the fuel and quell the flames.

Depending on a fire's location, firefighters and their equipment may be driven to or near a fire. Alternatively, firefighters can be inserted by air into an area near a fire to begin building a fire line.

Airtankers, scoopers, helicopters, or a combination of these aircraft can complement ground firefighting resources by dropping retardant, suppressant, or water on burning and potential fuel. Airtankers drop retardant, while helicopters and scoopers typically drop water, sometimes supplemented with suppressing foams or gels.

Fire retardants contain salts (typically fertilizers, such as ammonium sulfate or ammonium phosphate) mixed into water and dropped onto fuels to help build a fire line. Retardants continue to be effective even after the water has evaporated. By contrast, foams and water enhancers (gels) are suppressants and are added to water to increase the retardant's effectiveness by improving the accuracy of the drop and adhesion to fuels (see U.S. Department of Agriculture, Forest Service, Wildland Fire Chemical Systems, 2011). Suppressants are no longer effective once the water has evaporated. Water alone is a suppressant, but is not a long-lasting retardant because fuels will return to a flammable state after the water dissipates and evaporates.

Doctrine suggests that aerial application of suppressants and retardants will be ineffective without ground support (see, e.g., National Interagency Aviation Council, 2009, p. 85; Plucinski, 2009). Instead, airtankers, scoopers, and helicopters normally work in conjunction with firefighters on the ground. While a retardant drop from an airtanker may create a strip of land that that is resistant to ignition, even a small gap in such a barrier could render the effort worthless if the fire managed to pass over, through, or around the aircraft-generated barrier. On-the-ground firefighters are therefore necessary to secure the fire line created by aviation assets.

At the same time, there is a widespread belief among fire aviation professionals that air attack can be useful in delaying fire growth, or even suppressing small fires, before ground resources arrive. As we discuss later, the true value of an air attack in advance of ground firefighting resources is one of the key uncertainties affecting the results of our analysis.

A primary advantage of airtankers is their ability to quickly travel long distances to assist on-the-ground firefighters. Airtankers, scoopers, and helicopters can provide surge capability to assist firefighters who are already at the scene, though helicopters fly more slowly and have a smaller range than fixed-wing airtankers and scoopers, and scooper air speeds are higher than those of helicopters but lower than those of airtankers. In addition, scoopers and helicopters require access to a water source while airtankers do not. When water sources are proximate to a fire, scoopers and helicopters can have the advantage of faster cycle times than airtankers. With fast enough cycle times, a scooper might drop many more gallons of water per hour on a fire than a larger airtanker could drop retardant.

Although large airtankers are routinely used to fight already-large fires, there is comparatively little research documenting their effectiveness against such fires.[1] Indeed, Finney, Grenfell, and McHugh (2009) note that billions of dollars are spent annually to contain large wildland fires, yet the factors contributing to suppression success are poorly understood. Plucinski et al. (2007) offer anecdotal evidence that aerial suppression has saved homes, and we interviewed experts who cited many examples of the efficacy of aerial suppression in large fire attack. They noted that aircraft can be used to steer or turn fires, to control popup fires ignited beyond the main blaze, and to hold fire progression along lines too remote or steep for ground resources to operate effec-

---

[1] We wish to disclaim that an absence of evidence is not the same as evidence of absence. Finney (2007) notes that the modeling of large fires is relatively new and very complex. Perhaps airtankers can be valuable against large fires, but modeling tools simply are not able to demonstrate this effect commensurably with how cost-effectiveness in initial attack can be modeled.

tively. However, others suggested that aerial attack is often used inefficiently or to satisfy public demand for demonstrable effort, leading to what Cart and Boxall (2008) describe as "CNN drops."

There is firmer evidence that aircraft can assist with initial attack, particularly by supporting on-the-ground firefighters in containing a potentially costly fire while it is still small. Given the costs associated with large fires (property damage, lives lost, firefighting expenditures, and forest resource changes), there can be considerable value in a successful initial attack. The National Interagency Aviation Council (2009, p. 99) indicates that airtankers are often reserved for initial attack because of their high cruise speed and long range.

Because the cost-effectiveness of aerial attack against already-large fires is largely unknown, our approach to estimating the fleet size and mix for optimizing returns on government investment involved dividing the problem. First, we estimated the optimal fleet for initial attack, basing this analysis on a wealth of available evidence and models.[2] Then, we developed an analytic framework for answering the less well-understood problem of how many additional aircraft might be justifiable under a range of plausible assumptions about the benefits of fighting already-large fires with aircraft.

Multiple state, local, and federal agencies have a stake in wildland firefighting. Not surprisingly, many government entities own or contract for firefighting aircraft. Many of these aircraft are small, e.g., Type II and Type III helicopters that can be used to transport firefighters and their equipment or water. The Forest Service reports somewhat more Type II helicopter hours for transporting personnel than for dropping water, with carrying cargo as a tertiary mission (U.S. Department of Agriculture, Forest Service, 1992). The DOI's Bureau of Land Management contracts for a number of single-engine airtankers that carry 800 gallons of retardant. The states of Alaska, California, and Oregon separately contract for their own airtankers (U.S. Department of Agriculture, Forest Service, and U.S. Department of the Interior, 1995), and California operates 23 1,200-gallon airtankers (Cali-

---

[2]   We use the term *optimal* in a fairly broad sense that captures questions of robustness (e.g., whether a given solution is sufficiently flexible to handle a set range of possible future states).

fornia Department of Forestry and Fire Protection, undated). CL-215 and CL-415 scoopers are operated extensively in Canada, as well as in Alaska, California, and Minnesota.

Our focus was on the number and mix of large aircraft that the Forest Service needs to optimize social returns on public investment. We restricted our analysis to *large aircraft*, a term we use to denote Type I helicopters and 1,500- to 3,000-gallon airtankers and scoopers. We were instructed not to consider very large airtankers, such as converted 747s or DC-10s that can carry 10,000–20,000 gallons of retardant. Nor did we consider single-engine airtankers, Type II or Type III helicopters, or other aircraft that carry fewer than 1,000 gallons of water or retardant.

Our analysis did not consider use of military-operated aircraft (e.g., Air Force C-130 cargo aircraft) in firefighting.

With the exception of state-contracted aircraft and occasional aerial suppression efforts by National Guard aircraft, most large aircraft used in aerial suppression in the United States have belonged to the Forest Service. The Forest Service contracts for these aircraft from private providers, which maintain and operate them. Commercial vendors offer the large aircraft through "exclusive-use" contracts for entire fire seasons or through "call-when-needed" contracts, offered on a daily spot market. Contracts traditionally have a per-day fee and a per-flying-hour fee. Call-when-needed contracts tend to be more costly on a daily basis, but they provide the Forest Service with flexibility and minimize its financial commitment in case a fire season is mild.

The current fleet of airtankers available to the Forest Service is very old. Today's Forest Service–contracted airtankers are P-2Vs, former military aircraft dating back to the 1950s. The contractors obtained these aircraft after many years of military usage, refurbished them to serve as airtankers, and now maintain and operate them. This fleet's age and the accidents discussed in Chapter One have led to calls for its replacement and, ultimately, to this research report.

The Costs of Large Fires

Our initial attack analysis had two key input parameters. The first, the subject of this chapter, was the costs of large fires, $C_F(f)$.[1] The second, the subject of the next chapter, was the cost of aviation, $C_A(a)$. Ultimately, our objective was to find the mix of aviation assets that would minimize the total social costs of wildfires, including cost of large fires and the cost of large aircraft, $C_F\big(f(a)\big) + C_A(a)$. Chapters Five and Six explore these options using the National Model and the Local Resources Model, respectively. If a large fire were quite costly, we would ascribe considerable value to successful initial attack, thereby possibly justifying the considerable use of aviation to prevent a small fire from becoming large.

Wildfire suppression expenditures in the United States have risen dramatically over the past decade, to an average of $1.65 billion annually. In contrast, in the three decades prior to 2000, these costs averaged below $450 million (Liang et al., 2008), as shown in Table 3.1.[2]

---

[1] In this chapter, our definition of a large fire is any fire that burns more than 100 acres. Note that the models in Chapters Five and Six do not, in fact, use this 100-acre threshold. Rather, both models calibrate whether or not containment occurs in the first day of an initial attack. Achieving containment in the first day is clearly correlated with the size of a fire. For the purposes of estimating what an escaped (not contained) fire costs, we used a 100-acre cutoff. This is an imperfect relationship, however. For example, a fire might be contained within a day but be larger than 100 acres. That said, we use the 100-acre cutoff as our approximation to map fire escapes in Chapters Five and Six and to develop average cost estimates for large fires in this chapter.

[2] Liang et al. (2008) actually cite a figure below $400 million in 2005 dollars. We have converted all dollar values in this report to 2010 terms using the U.S. Department of Commerce (2010) gross domestic product price deflator.

Table 3.1
Annual Federal Wildfire Suppression Expenditures, 1997–2008
(2010 $ millions)

Year	Forest Service	DOI Agencies	Total
1997	234.4	137.5	371.8
1998	397.2	142.3	539.5
1999	460.5	196.9	657.5
2000	1,344.4	418.3	1,762.7
2001	834.7	329.4	1,164.1
2002	1,537.3	474.8	2,012.1
2003	1,204.1	357.2	1,561.3
2004	830.3	321.6	1,151.9
2005	763.8	325.5	1,089.3
2006	1,608.5	454.4	2,063.0
2007	1,457.9	478.0	1,935.9
2008	1,486.8	400.4	1,887.2

SOURCE: National Interagency Fire Center (NIFC), personal
communication, June 7, 2010.

Indeed, every year between 2000 and 2010 saw more than $1 billion
in federal suppression expenditures, including 2002 and 2006, when
expenditures exceeded $2 billion.

High federal suppression costs have triggered concern and calls
for action by Congress and the public, but they represent only the most
easily quantified economic costs of large wildfires. In addition to these
federal costs, there are large state and local suppression and emergency
management expenditures, as well as business losses (and gains) associated with large wildfires. Fires destroy homes, infrastructure, and
culturally significant sites. Smoke from fires can cause severe or life-
threatening respiratory ailments. Firefighters die in the line of duty.
Fires can degrade the resource values provided by forests and wildlands
by, for instance, destroying timber, grazing lands, and crops; damag-

ing watersheds and the habitats of endangered species; releasing large quantities of carbon into the atmosphere; and eroding and destabilizing soil.

But wildfires do not solely produce social and environmental costs. They can also provide a range of ecological benefits, such as maintaining native fire-adapted flora and fauna, reducing pest infestations, increasing water supplies, enriching soils, and limiting the spread of invasive plant species. The net (positive or negative) ecological ledger on wildfires is difficult to determine, either for a specific fire or more generally.

Fires consume forest fuels that would otherwise remain at risk of burning later. As such, fires can have a role in decreasing future fire sizes and, hence, costs. Conversely, successful fire suppression contributes to the accumulation of fuel. Indeed, more than half of the wildlands managed by the Forest Service have gone two or more times as long without burning as they did before aggressive suppression policies were implemented in the early 20th century. These policies have led to long-term changes in the accumulation and distribution of fuels that contribute to more intense, faster-moving, and larger wildfires (Mutch, 1994; Hesseln and Rideout, 1999). Some of the surge in wildland fire costs reflected in Table 3.1 may be attributable to this increasing intensity of large fires.

In estimating the cost of large fires, it is necessary to tabulate the available data on federal suppression expenditures, the state and local suppression costs of fires to which the Forest Service has responded, federal post-fire rehabilitation expenditures, insured losses, fatalities, and future suppression expenditure savings.

Our tabulation did not include state and local suppression costs for fires to which the Forest Service did not respond, state and local emergency management costs, the nonmarket value of changes to ecosystems, recreation services, carbon dioxide release, species habitats, timber resource changes, water availability and purity, the public health effects of smoke, or unreimbursed individual and volunteer expenses. These nonmarket effects could be quite large.

Given the uncertainty of the impact of these nonmonetized costs on our results, we evaluated the sensitivity of our air fleet optimization models using a wide range of possible values for the average cost of a large fire. Thus, although our estimate of the average cost of a large fire ranges from $2.1 million to $4.5 million, we also tested a much wider range of values, from $300,000 to $10 million, as discussed in Chapter Six (see Table 6.2).

Costs and Benefits of Wildland Fire

Forest Service aviation assets are deployed to fires on Forest Service and DOI land, as well as to state fires and to other fires on an as-requested basis, with priority given to initial attack. Although federal wildland firefighting costs are well documented, many associated federal, state, and local costs are not. For instance, we are not aware of authoritative estimates of the suppression costs incurred by state and local governments. Although a few studies have sought to estimate these costs for individual large fires (see, e.g., Dunn, 2003; Graham, 2003; Morton et al., 2003), the special circumstances that made those fires worthy of investigation also make them unsuitable for drawing generalizations about state and local suppression expenditures. Neither are national estimates of the local costs for emergency response activities, such as evacuations and road closures, available. As shown in Table 3.2, the vast majority of wildfires are on state, local, or private land, suggesting that state and local costs might dominate national suppression and emergency management expenditures.

To address the omission of state and local expenditures, we estimated per-fire costs for the Forest Service and the agencies with fire management responsibilities in DOI (the Fish and Wildlife Service, the National Parks Service, the Bureau of Land Management, and the Bureau of Indian Affairs). Because Forest Service and DOI costs per acre span a wide range, we adopted the assumption that state and local fire suppression costs per fire fall somewhere between the Forest Service and DOI values.[3]

---

[3]   Cal Gale of the NIFC told us that Forest Service fires are frequently more costly because of the fuel types and topography involved. Such timber types require that fire crews spend a

Table 3.2
Number of Wildfires and Annual Acres Burned, by Agency

Year	Number of Wildfires				Acres Burned			
	Forest Service	DOI	State/Other	Total	Forest Service	DOI	State/Other	Total
1999	10,424	7,443	75,835	93,702	717,679	3,059,609	1,884,688	5,661,976
2000	11,699	8,865	71,716	92,280	2,333,672	2,549,219	2,510,602	7,393,493
2001	10,717	9,075	64,204	83,996	595,268	1,283,214	1,691,743	3,570,225
2002	9,246	8,100	56,077	73,423	2,402,501	2,287,066	2,493,412	7,182,979
2003	10,250	7,862	45,156	63,268	1,428,266	1,144,536	1,386,420	3,959,222
2004	8,608	7,440	49,413	65,461	551,966	3,515,841	4,030,073	8,097,880
2005	7,331	8,695	50,727	66,753	781,148	5,757,416	2,150,825	8,689,389
2006	10,403	11,677	74,305	96,385	1,896,071	3,093,758	4,883,916	9,873,745
2007	8,486	8,091	69,128	85,705	2,835,577	2,891,099	3,601,369	9,328,045
2008	7,113	7,696	64,140	78,949	1,234,479	684,330	3,373,659	5,292,468
2009	7,691	7,794	63,307	78,792	715,677	2,193,476	3,012,633	5,921,786

SOURCE: National Interagency Coordination Center, 2009.

Summary of High and Low Estimates of the Cost of Large Fires

Because there are considerable uncertainties in quantifying the costs and benefits of large fires, we produced separate high and low estimates for each of several quantifiable components of large fire costs. Table 3.3 summarizes these component estimates. By systematically selecting assumptions that lead to either especially high or especially low estimates for the average cost of a large fire, sought to bracket the range of fleet mix options that could be justified on the basis of the estimable social costs of large fires. Our approach is discussed in the next section, with a more formal calculation for each offered in Appendix A.

Table 3.3
High and Low Estimates of the Average Cost of a
Large Fire (2010 $ thousands)

Cost Category	Low	High
Fire suppression		
Federal	1,879	3,300
State/local	132	231
Large aircraft costs	(250)	0
Small fire cost	(7)	(7)
Rehabilitation	50	50
Insured losses	329	769
Lives	61	127
Future suppression	(76)	(19)
Total	2,117	4,451

NOTE: Amounts in parentheses are cost savings.

---

higher percentage of their time on labor-intensive mop-up to adequately contain the ignition. A larger percentage of DOI lands are shrub or grass with generally less steep topography, resulting in less time required for mop-up and reducing the per-acre cost.

One should view Table 3.3's values as estimates of the average cost of an additional large fire,

$$\frac{\partial C_F(f)}{\partial f}.$$

Averaging the low and high estimate yields an average cost per large fire of approximately $3.3 million.

Summary of the Costs and Benefits of Wildfires Included in the RAND Analysis

Federal Fire Suppression Costs for Large Fires

A disproportionate share of federal fire suppression spending is allocated to fighting large fires. Calkin, Gebert, et al. (2005) report that large fires represented only 1.1 percent of all fires between 1970 and 2002 but accounted for 97.5 percent of acres burned.

We consider a fire to be large if it involves 100 or more acres. With this definition, Table 3.4 shows that, between 2005 and 2009, 6.5 percent of all federal fires were large, but they consumed 92.9 percent of federal suppression expenditures.

Overall, federal costs have averaged about $1.2 million per large fire, ranging from an average of $150,000 for DOI large fires to an average of $3.6 million for Forest Service large fires.

These NIFC data are derived from fire manager cost estimates that may overestimate true suppression costs. A more conservative estimate was put forward by Gebert, Calkin, and Yoder (2007), who used data from Forest Service financial accounting systems. They found that, between 1995 and 2004, an average Forest Service costs ran about $1.5 million per large fire, or $1,136 per acre burned. Continuing collection of the same administrative data shows that, between 2005 and 2009, these costs rose to $3.6 million per Forest Service fire exceeding 300 acres (Gebert, 2011). Extrapolating this number to our definition of large fires, we estimated that Forest Service fires covering more than

Table 3.4
Federal Fire Costs by Fire Size and Agency, 2005–2009 (2010 $ millions)

Fire Size Class	Forest Service		DOI		Total	
	Number	Cost	Number	Cost	Number	Cost
A: 0–0.25 acres	26,605	53.7	15,150	43.5	41,755	97.2
B: 0.25–10 acres	10,518	118.3	13,769	72.6	24,287	190.8
C: 10–100 acres	2,394	112.4	4,278	72.0	6,672	184.4
D: 100–300 acres	563	99.6	1,297	40.6	1,860	140.2
E: 300–1,000 acres	373	605.5	999	77.8	1,372	683.2
F: 1,000–5,000 acres	362	1,528.2	741	125.7	1,103	1,653.9
G: More than 5,000 acres	244	3,394.9	513	287.7	757	3,682.6
All fires	41,059	5,912.6	36,757	719.8	77,806	6,632.4
Large fires	1,542	5,628.2	3,550	531.7	5,092	6,159.9

SOURCES: NIFC and FIRESTAT data.
NOTE: FIRESTAT is the Forest Service's database of historical fire statistics. It contains vital statistics about wildfires occurring on Forest Service land since 1970. Statistics such as cause, location, cost, response equipment, and weather are recorded by forest managers and federal firefighters.

100 acres cost $2.3 million per fire (lower than the $3.6 million estimate derived from Table 3.4).

Using these high and low estimates for the suppression costs of fighting Forest Service fires, we must also adjust for the fact that some portion of fires fought by large aircraft may be the comparatively inexpensive DOI fires. According to managers at the Forest Service's fire and aviation management office, 80–90 percent of large airtanker efforts are directed to Forest Service fires rather than comparatively inexpensive DOI fires. We used this 80–90 percent estimate, along with NIFC and Forest Service financial data, to produce high and low weighted cost estimates for large wildfires. Specifically, for our high cost estimate, we used a weighted average cost consisting of 90 percent of the Forest Service cost estimate of $3.6 million (shown in Table 3.4) and 10 percent of the DOI large fire cost estimate of $150,000 (also shown in Table 3.4) for a total of $3.3 million per large fire. For our low cost estimate, we used a weighted average of 80 percent of the more conservative Gebert-derived extrapolation of $2.3 million and 20 percent of the DOI cost estimate of $150,000 (shown in Table 3.4) for a total of $1.9 million per large fire.[4]

State and Local Fire Suppression Costs

There have been several studies of suppression and emergency management costs borne by state and local governments for individual fires. Examining the 2003 Old, Grand Prix, and Padua fires, for instance, Dunn (2003) calculated that California state and county expenses amounted to more than $18 million, or more than 40 percent of the costs incurred by the Forest Service against those fires. Further, the $18 million estimate ignored costs borne by local municipal fire departments.

Estimates of total annual cost to states and localities of fire suppression and emergency management are unavailable. However, Gebert and Schuster (2008), examining suppression expenditures in the Southwest in 1996 and 1997, found that Forest Service expenditures amounted to 87 percent of total costs, with state and local costs accounting for 7 percent of the total. Using this estimate, we assumed that state and local suppression costs accounted for 7 percent of our two different federal fire suppression cost estimates—so, $142,000 per large fire in the low case and $248,000 per large fire in the high case.

---

[4]   We did not find any evidence that Category B fires (those against which aviation assets are most effective) have different suppression costs (conditional on becoming large) than the faster-rate-of-spread fires that aviation cannot contain (Category C fires). We examined the estimated costs of all fires that burned more than 100 acres of Forest Service land between 2004 and 2009 as reported in the FIRESTAT database. The estimated Forest Service costs of large Category A and B fires, $2.3 million, was not much different from (and, indeed, was greater than) that of Category C fires, $2.2 million. Because we did not see evidence that our Category B fires were less costly than our Category C fires, we used the average costs per fire between 2004 and 2009 (shown in Table 3.4) to estimate foregone federal suppression expenditures when large aircraft prevented a fire from becoming large.

Large Aircraft Costs

A key rationale for employing large aircraft is to prevent the large fires that account for a disproportionate share of total wildfire management costs. But a portion of the high cost of large fires is attributable to the use of large aircraft against them. Ideally, we would estimate the expected cost of large fires given various fleet mix options, but we lacked sufficient evidence to determine whether any number of large aircraft would increase or decrease total large fire costs. Instead, we considered fire costs with large aircraft in our high cost estimate and without large aircraft in our low cost estimate.

Brosnan (2008) reported annual costs of $93 million for the existing fleet of 19 large airtankers and $114 million for large helicopters. These costs make up 83 percent of the Forest Service's total aviation budget, with the remainder covering smaller aircraft and the infrastructure costs of maintaining and staffing aircraft bases. The proportion of aircraft operations directed against large fires is not well documented. The U.S. Government Accountability Office (Dalton, 2009) suggested that up to one-third of the cost of fighting large fires may be attributable to aviation costs, beyond those associated with large aircraft. By contrast, the Large-Cost Fire Independent Review Panel (2009) and the Independent Large Wildfire Cost Panel (2008) reported aviation assets accounted for 18 percent of total costs in 2007 and 14 percent of total costs in 2008. We therefore estimated that 16 percent of large fire costs come from aviation. Because large aircraft account for 83 percent of Forest Service aviation costs, we estimated that large aircraft account for 83 percent of 16 percent of large fire costs. In our low estimate of large fire costs, we therefore deducted about $250,000 (83 percent of 16 percent of $1.9 million) from the suppression costs to eliminate large aircraft costs from the total.

Small Fire Costs

We further adjusted our estimate of the cost of large fires to account for the fact that prevention of a large fire nevertheless incurs the cost of a small fire, so the entirety of large fire suppression costs cannot be counted as benefits of a successful initial attack. In Table 3.3, the average Forest Service suppression cost for a small fire is about $7,000.

Rehabilitation of Burned Lands

Immediately after a fire, rehabilitation efforts are often required to stabilize and prevent the erosion of soil, to repair public infrastructure (such as roads and culverts), and to protect watersheds and habitats. Some of these costs, such as those for Burned Area Restoration and Evaluation teams, are included in the Forest Service's suppression costs. In addition, Forest Service annual budget documents report that, between 2005 and 2009, the agency spent $77 million on fire rehabilitation and restoration through regular and emergency appropriations (U.S. Department of Agriculture, Forest Service, 2005, 2007, 2009, 2010). We assume that the majority of these funds were directed to the 1,542 large fires over this period for a per-fire rehabilitation cost of about $50,000. This is undoubtedly a low estimate, however, since some individual fires cost more than this. For example, Lynch (2004) reported that the Forest Service spent $25 million on emergency rehabilitation after the 2002 Hayman fire in Colorado. Further, this total does not include significant rehabilitation costs incurred by volunteers, nongovernmental organizations, and other public and private agencies.

Insured Losses

The spread of homes into the wildland-urban interface (WUI) is a widely cited factor contributing to the recent increase in fire suppression costs. The tripling of new housing units between 1940 and 2000 has disproportionately occurred in non-metropolitan counties as suburbs and exurbs have attracted more homeowners to the WUI. Hammer, Stewart, and Radeloff (2009) estimated that, by 2000, 38 percent of all housing units and 11 percent of all land area in the 48 contiguous states fell within federally defined WUI areas.

Housing growth in the WUI leads to predictable losses of private homes. Gude et al. (2009) found that approximately 10,000 homes were lost to wildfires between 2002 and 2006. Using an estimated average replacement cost for homes in western states of $193,374 for the years 2005 to 2010 (Davis and Heathcote, 2007), this figure translates to annual wildfire-related housing losses of approximately $387 million. Comparably, the Insurance Information Institute (undated) has estimated that wildfires account for $484 million in annual insured property claims. A markedly higher estimate is provided by the International Code Council (2008), which suggested that 21,800 structures were lost to wildfires between 2000 and 2007. These lost structures resulted in annual claims of $905 million, or about $374,000 per structure.

For our low and high estimates of damage to houses and structures attributable to large wildfires, we divided the total annual estimates (low of $387 million extrapolated from estimates by Gude et al., 2009, and Davis and Heathcote, 2007, and high of $905 million from the International Code Council, 2008) by NIFC estimates of the average annual number of federal, state, and other large wildfires in the United States between 2004 and 2009—1,177—resulting in structural losses per large fire ranging from $329,000 to $769,000.

Loss of Life

Firefighting is risky. Of 173 wildland firefighter fatalities between 1999 and 2006, most were the result of large fire accidents (National Wildfire Coordinating Group, 2007; Large-Cost Fire Independent Review Panel, 2009). Because precise information on how many fatalities occur in the course of fighting small fires is not available (see Abt, Prestemon, and Gebert, 2008), we assume that fatalities are distributed across fires in proportion to the fires' size. Using statistics on federal wildfire sizes, we estimated that 88 percent of fatalities were attributable to large fires, giving us an average of 19.03 fatalities per year from large fires. Again, assuming that there are 1,177 state and federal large fires per year (National Interagency Coordination Center, 2005–2009), this equates to an expected 0.0162 deaths per large wildfire.

Government regulatory cost-benefit analyses routinely estimate the value of a statistical life saved. Recent Environmental Protection Agency guidelines recommend using the value of $7.87 million, whereas the Federal Aviation Administration (FAA) has suggested using $3.75 million (Viscusi and Aldy, 2003). Therefore, we used $61,000 as our low expected total fatality cost ($3.75 million multiplied by 0.0162) and $127,000 as our high expected total fatality cost ($7.87 million multiplied by 0.0162).

Aircraft accidents are among the leading causes of firefighting deaths (National Wildfire Coordinating Group, 2007). The Blue Ribbon Panel on Aerial Firefighting (2002) reported that, historically, about one large airtanker is lost to in-flight accidents per year, with many of these incidents causing fatalities. From 2000 to 2009, NIFC fatality reports show 12 airtanker fatalities, 33 helicopter fatalities, and another nine unspecified aircraft fatalities.

Although we hope that the rates of fatalities associated with wildfire aviation will decrease with the acquisition of new aircraft, we have not tried to account for any possible future decrease in expected aviation fatalities per large fire.

Future Suppression

Large fires are costly to fight but might produce savings in later years if the areas they burn enjoy a period of reduced fire risk or reductions in fire severity. Conversely, the successful initial attack of fires might merely postpone large fires, even increasing their severity by allowing additional fuel to accumulate.

The effects of prescribed burns and past wildfires on fire behavior are not as straightforward as might be expected. For instance, according to Brown and Davis (1973, pp. 318–391), in some vegetation classes, burns create

> dangerous volumes of dead timber over extensive areas constituting an intolerable hazard. It is a common saying that one wildfire breeds another in coniferous forests. . . . Much of the area burned by the Tillamook fire of 1933 disastrously burned again in 1939, 1945, and 1951. . . . Instances of fire-killed timber resulting from the residue of one wildfire helping to trigger a second and sometimes a third fire can be cited almost endlessly.

Similarly, detailed analyses of the effects of past fires and prescribed burns on the Hayman fire concluded that, with the exception of fires that occurred within one year of a prior fire, burned areas from past fires did not stop new fires. Instead, they moderated the severity of the fire by, for instance, preventing the spread of crown fires, the dangerous fires that spread across the forest canopy (see Martinson, Omi, and Shepperd, 2003).

Attempts to quantify the effects of burned areas on future wildfires suggest that modest reductions in fire severity may be found for eight to 12 years for a variety of fuel types. In a systematic review of multiple conifer forest burn areas that either had or had not already burned in the prior ten years, Martinson and Omi (2003) found that scorching in the tree crowns of previously burned areas was 60 percent less severe than that in unburned areas. Similarly, a statistical analysis of Florida fires by Mercer et al. (2007) found that for every percentage increase in prescribed burn area, there is a short-term (two-year) 0.71-percent reduction in severity-weighted wildfire acres.

More often, fire models are used to estimate the effects of burns or other pretreatments on later wildfire behavior. For instance, after modeling fire severity in Arizona experimental forest areas, some of which had been exposed to historically expected fire frequencies and some of which had not, Fulé et al. (2001) found 43- to 70-percent reductions associated with recent fires on five measures of fire severity (crown percentage burned, rate of fire spread, heat/area, flame length, and torching index) and an 18 percent reduction on a sixth measure (crowning index).

In our tabulation, we made two important assumptions about the effects of past fires on later fire suppression costs. The first was that burned areas do not prevent later fires. Instead, we assumed that burns reduce the intensity of later fires by 30–70 percent (for our low and high estimates) for three to 12 years (again, for our low and high estimates). These estimates are roughly consistent with the few studies that have attempted to quantify the effects of wildfire and prescription fires on later fire severity. Although these broad assumptions ignore differences in fuels, moisture, canopy, weather, and other factors that undoubtedly have dramatic effects on the true effects of prior burns, we judged our parameter ranges to be a reasonable simplification for our purposes.

A second important assumption is that firefighting costs will be proportional to fire intensity, so a 50-percent reduction in fire severity can be treated as a 50-percent reduction in suppression effort and cost.

Rideout and Ziesler (2008) argue that it is theoretically possible that pretreatments, like planned burns, actually increase the suppression effort that a region might require to minimize social costs. Conversely, the lower fire severity expected in pretreated areas might reduce the chances of large fires, not just their severity—meaning that suppression costs might be reduced by a much larger factor than severity reductions. Nevertheless, our simplifying assumption is supported by observations that fire severity is closely linked to fire size, which, in turn, is generally proportional to suppression expenses, at least across large wildfires (Calkin, Gebert, et al., 2005; Gebert, Calkin, and Yoder, 2007).

Appendix A provides an illustration of how we developed our future suppression calculation. We ended up with estimates of future cost savings resulting from a previous large fire that modestly reduce the cost of a future large fire. Specifically, we found that large fires might have a net present value for future fire suppression savings of between $19,000 and $76,000.

Costs and Benefits of Wildfires Excluded from the RAND Analysis

There are several important costs and resource value changes that we did not include in our analysis because the data did not exist or because the data or methods required to estimate the values were not yet sufficiently well developed to justify including them. These included nonmarket values, federal disaster assistance, timber losses, and public health effects.

Nonmarket Values

There is a growing literature on how fires affect nonmarket resources, such as wildlife habitats, watersheds, public health, cultural heritage, and recreational services. But the data and methods required to quantify these resource changes are not yet adequate to generate compelling estimates of these nonmarket values (Abt, Prestemon, and Gebert, 2008; Hesseln and Rideout, 1999; Venn and Calkin, 2007).

Results from the small number of studies that have attempted such valuations reveal complex effects of wildfires that may correspond to large positive or negative value changes and to changes that evolve dynamically over time. For instance, Englin, Holmes, and Lutz (2008) reported complex time-varying effects from fires on recreational demand, with demand increasing in the years immediately following a fire but significantly decreasing in later decades.

Similarly, some fire effects result in both positive and negative changes, such as the effects of fire on watersheds. Typically, fires increase water yield, but they also increase sedimentation in the water. Potts, Peterson, and Zurring (1985) attempted to value these positive and negative outcomes, finding that the benefits of additional water availability exceeded the costs of additional sedimentation by a factor of more than 1,000 in some regions.

Other studies have suggested that the social value of preventing large fires may be extremely high. For instance, using a contingent valuation approach to estimating willingness to pay, Loomis and Gonzales-Caban (1998) found that the societal value of protecting the first 1,000 acres of northern spotted owl habitat in California and Oregon amounted to $25 per household, a figure that Venn and Calkin (2007) note is "greater than the annual national fire suppression expenditure by the Forest Service in recent high cost firefighting years."

The uncertainties in nonmarket values have led recent Forest Service guidance on cost-risk analyses to suggest estimating the minimum value of nonmarket effects that would be implied by available interventions, rather than trying to estimate the nonmarket effects themselves (Calkin, Hyde, et al., 2007).

Federal Disaster Assistance

In principle, federal disaster assistance funding to states, localities, and individuals affected by wildfires should be known to the federal government. However, a recent Congressional Research Service report on wildfires reported that public data on Federal Emergency Management Agency fire disaster assistance were not available (Gorte, 2006).

A portion of this funding goes to reimbursing states and local governments for the costs they incur in suppressing wildfires. Because we have already estimated state and local suppression costs, including the costs of federal reimbursements in our calculations would result in double-counting of some suppression costs. Therefore, this portion of the federal disaster assistance budget is omitted from the analysis.

Timber Losses

While past studies have attributed considerable financial value to lost timber (e.g., 20 percent of total losses for four large fires reviewed by Abt, Prestemon, and Gebert, 2008), such estimates reveal widely varying assumptions about the commercial viability of the lost timber, the proportion of existing timber lost in burned areas, and the value of salvageable burned timber. Some estimates of timber losses consider only Forest Service timber lease values or lost Forest Service timber sales, while others have considered the likely wealth transfers resulting from the market effects of shocks to the timber supply.

Butry et al. (2001), for instance, examined the welfare effects of the timber price reductions resulting from a short-term glut of salvaged timber and later price increases resulting from local shortages of timber, which they hypothesized would follow the 1998 wildfires in northeast Florida. Their analysis suggested that short-term gains to consumers are roughly matched by long-term gains to producers, with a net loss resulting from the effects on owners of damaged timber.

We adopted a social cost perspective for our analyses, as opposed to considering just the cost to the Forest Service. As such, to include timber losses in our analysis, we would have required an estimate of the welfare changes resulting from timber losses that are net of the types of wealth transfers described by Butry et al. (2001). While that study produced such an estimate for one fire in 1998, we do not believe that these results can be generalized to timber losses nationally, and other, more general estimates of timber resource value changes of this kind were not available for us to incorporate into our analysis.

Public Health Effects

Fires degrade air quality in ways that are harmful and can exacerbate asthma and bronchitis, reducing quality of life, increasing hospital admissions, and contributing to deaths. But attributing the prevention of morbidity and mortality to fire suppression is complicated by uncertainties and great variation in the numbers of people affected by individual fires, the severity of harms they might be expected to suffer, and the valuation of those harms.

Sorenson et al. (1999) noted that, during the 1998 Florida wildfires, admissions at some regional hospitals increased 91 percent for asthma and 132 percent for bronchitis over the same period in the prior year, though the atmospheric conditions that contributed to fire risk in 1998 could have affected respiratory conditions as well. Butry et al. (2001) used treatment costs as a proxy for the societal costs of smoke from the same Florida fires, concluding that they represented a small cost relative to other costs of the fires. Similarly, Rittmaster et al. (2006) evaluated the health effects of a large fire in Canada. Unfortunately, the results of each of these studies are likely unique to the fires they investigated. We know of no analyses offering national or per-acre-burned estimates of the public health costs of smoke, so we could not include these values in our analysis.

Next, we discuss the estimated costs of prospective aircraft. Ultimately, it is these aircraft costs against which we compared the estimated costs of large fires.

CHAPTER FOUR

The Costs of Large Aircraft

Chapter Three estimated the costs of large fires and, hence, the benefits of successful initial attack. This chapter focuses on the life-cycle costs of large aircraft, $C_A(a)$. Specifically, we estimated annualized aircraft costs or the expected annual payment associated with operating each of several types of large aircraft. Our estimates are broad enough to represent the expected costs under a range of ownership structures, including aircraft owned and contracted by the Forest Service. This analysis does not make recommendations about whether the Forest Service should own or contract for its firefighting aircraft.

Per the Forest Service's direction, the RAND research team did not directly interact with aircraft manufacturers to learn about aircraft acquisition or operating costs. Instead, we restricted our study to publicly available information. Our cost estimates are therefore not "source selection–quality." A Forest Service source selection would require bids from competing contractors, generating a degree of cost-estimation fidelity that we cannot provide in this chapter.

Candidate Aircraft

The primary large airtanker in use today is the P-2V. It is a former military aircraft that was converted for use in firefighting. It carries 2,082 gallons of retardant.[1] These are contractor-owned, contractor-operated

---

[1] This aircraft previously carried 2,450 gallons, but its capacity was downgraded as a safety measure following the 2002 crashes.

aircraft under contract to the Forest Service to provide support in fighting wildfires. Similarly, the Forest Service contracts for helicopter services from private owner/operators. We selected representative candidate replacement aircraft based on three criteria. First, we focused on Type I helicopters and 1,500- to 3,000-gallon airtankers and scoopers. At the Forest Service's direction, we did not examine smaller single-engine airtankers that carry 800–900 gallons of retardant or very large airtankers that carry 10,000–20,000 gallons of retardant.

Second, we focused on aircraft that will be available within the next few years. This restriction limited us to aircraft currently in production, since it may not be possible to design and build a new aircraft in the necessary time frame. Further, a new design would probably be cost-prohibitive, given the high development costs for a new design and the small quantity that the Forest Service is likely to require (perhaps 15–30 aircraft).

Finally, we limited the candidates to aircraft that have established training and logistical support systems in North America. The relatively small size of the Forest Service fleet could make establishing new training and logistical infrastructure cost-prohibitive. In contrast, if an aircraft already in service with the U.S. government is chosen, the Forest Service could leverage an existing infrastructure and limit the need for new training and support activities. For example, if the C-130J-30 were chosen, the Forest Service could have its pilots or its contractors' pilots train on existing aircraft at Air Force training bases, avoiding the need to create a dedicated training activity.

The Forest Service asked the RAND team not to analyze specific manufacturers' products, but instead to select representative types of aircraft spanning a range of sizes and capabilities. We analyzed five candidate aircraft drawn from three categories: 1,500- and 3,000-gallon airtankers, a 1,600-gallon scooper, and 1,200- and 2,700-gallon helicopters. These candidate aircraft cover the range of large firefighting aircraft in use today.

To represent the 1,500- and 3,000-gallon fixed-wing airtankers, we chose the Alenia C-27J and the Lockheed Martin C-130J-30, respectively. A C-130J-30 can be converted to an airtanker role with the addition of a roll-on retardant tank system. Such a conversion can be accomplished in less than four hours.[2] When the retardant system is removed, the aircraft retains its full cargo-carrying capacity. Another option would be to develop an integrated tanking system that would be part of the aircraft's baseline equipment. However, we used the existing removable C-130J-30 tanking system for our performance and cost estimates.[3]

We also examined a commercial-derivative fixed-wing aircraft with a 3,000-gallon capacity. Retardant tanks, pumps, and nozzles would have to be integrated onto these aircraft to allow them to operate as airtankers. We were unable to identify the acquisition cost of commercial aircraft because of uncertainty about aircraft pricing. We obtained published price lists for this type of aircraft, but commercial aircraft typically sell at a considerable discount from list prices. The actual price paid for an aircraft would be a function of negotiations between the Forest Service and the manufacturer or the outcome of a competitive bidding process. Or, if the Forest Service contracted for the aircraft, acquisition costs would implicitly be built into the annual contract cost.

The Forest Service indicated to us that it was reluctant to use a commercial-derivative airtanker. It also provided us with the following requirements for airtankers that it acquires:

- purpose-built for the airtanker mission or missions similar in terms of maneuver loads and low-level flight
- original equipment manufacturer support for the make/model, including maintenance, parts, and engineering support, for the life of the aircraft
- FAA-approved maintenance and inspection program for the aircraft's use as an airtanker and FAA-type certification for the aerial dispersal of liquids.

---

[2] See slide 20 in U.S. Department of Agriculture, Forest Service (2004).

[3] We were not able to obtain details on the conversion time that would be associated with a 1,500-gallon airtanker variant of the C-27J.

We also examined a scooper with a 1,600-gallon capacity. We used information on the Bombardier CL-415 to represent this type of aircraft. Scoopers generally carry water rather than retardant, typically picking up water from lakes and rivers rather than at an airport. Foam is often injected into the water to increase its efficacy. This type of aircraft combines some of the advantages of fixed-wing aircraft, such as speed, with the shorter cycle times associated with rotary-wing aircraft.

Finally, we examined two rotary-wing aircraft, the first with a 1,200-gallon payload and the second with a 2,700-gallon payload. Such helicopters generally carry water. The Forest Service has extensive experience with contracting for such helicopters. Our primary interest in helicopters was to estimate the extent to which they might cost-effectively replace airtankers and scoopers in initial attack. Helicopters are also used extensively in campaigns against already-large fires.

Cost Assessment Overview

We developed cost estimates for candidate aircraft based on publicly available information. For each candidate aircraft, we estimated the constant-dollar life-cycle costs (covering development, procurement, and operating and support, or O&S) and then translated those life-cycle costs into annualized or annuitized values. Table 4.1 summarizes our cost estimates.[4] These estimates do not include retardant or foam costs.

Suppose, for instance, that a candidate aircraft had estimated acquisition (development and procurement) costs Q, was expected to operate for 30 years, and would have annual constant-dollar operating costs O_t for those 30 years (with t as the year subscript, running from 1 to 30). We assume, for simplicity, that annual inflation-adjusted

---

[4]   Table 4.1 presents our best estimate of what the annualized costs—of ownership or leases—would be. The argument is that, in equilibrium, the Forest Service would pay roughly the same amount whether it owned the aircraft or leased it, since, in the latter case, ownership costs would be borne by the lessor, who would pass those costs to the Forest Service as the customer.

Table 4.1
Estimated Per-Aircraft Annualized Costs (2010 $ millions)

Aircraft	Development	Procurement	Annualized Acquisition	O&S	Contract	Annualized Total
1,600-gallon scooper		30.4	1.5	1.3		2.8
1,500-gallon military airtanker		51.1	2.5	2.7		5.2
3,000-gallon military airtanker		81.2	4.0	4.2		8.2
3,000-gallon commercial airtanker	2.5	85.0	4.3	2.8		7.1
1,200-gallon helicopter		31.2	1.5	1.7		3.3
2,700-gallon helicopter					7.1	7.1

NOTE: The 3,000-gallon aircraft commercial development cost is based on a $50 million program and 20 aircraft.

operating costs are constant over time. We translate the acquisition costs into an annualized cost, x, solving the equation

$$\sum_{t=1}^{30} \frac{x}{1.027^t} = Q,$$

where 2.7 percent is OMB's 2010-prescribed long-term real interest rate, and x corresponds to the "Annualized Acquisition" column in Table 4.1. Annualized total costs would be $x + O_t$. As noted, the annualized total cost could be viewed as the annual cost of aircraft ownership or as the annual payment to an aircraft lessor.

Data Sources

We drew our acquisition cost estimates for military-derivative aircraft from Air Force budget documents. Acquisition costs for commercial aircraft and the scooper were from list prices published in industry journals. We adjusted these costs to account for list price discounts (in the case of commercial aircraft), unneeded equipment, and modifications.

We generated the O&S cost estimates using data from Conklin & DeDecker (2009a, 2009b, 2010), a leading publisher of O&S cost data for commercial aircraft. When data for a desired aircraft were not available, we used an aircraft of similar size, weight, and configuration as a proxy. Data on O&S costs of military-derivative aircraft were drawn from the Air Force Total Ownership Cost System.

Traditionally, Forest Service contracts for airtankers have run from 140 to 180 days per year. The average number of flight hours per airtanker has typically ranged from 150 to 360 annually over the past several years. Type I helicopter contracts have typically run for 90 to 180 days, with these aircraft being flown 200–350 hours per year. We therefore assumed that new aircraft would operate for 140 days per year. Given that we are modeling new aircraft that should have fewer maintenance challenges, we assumed an increase in flying hours to 400 per year to generate our estimates of O&S costs.

We also assumed that new aircraft would have a 30-year service life, that fuel would cost $4 per gallon, that a fixed-wing air crew is composed of two pilots and one flight engineer, and that there would be 1.1 crews per fixed-wing aircraft and 1.5 crews per helicopter. We also assumed that ground crews would consist of two refuelers and two mechanics per aircraft.

Airtanker and Scooper Cost Estimates

As shown in Figure 4.1, we estimated that annualized fixed-wing life-cycle costs per aircraft would range from $3 million to $8 million without retardant or foam costs. Not surprisingly, larger airtankers have greater annualized costs.

Figure 4.1
Estimated Annualized Fixed-Wing Airtanker Costs, Without Retardant or Foam

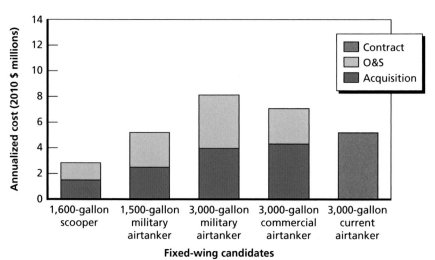

RAND *MG1234-4.1*

The 1,600-gallon scooper is the least expensive candidate aircraft on a per-aircraft basis. Aircraft costs are highly correlated with weight, and it is the lightest of the candidates. The 1,500-gallon military-derivative airtanker is more expensive than the slightly higher-capacity scooper because of the nature of its design. The scooper is purpose-built for carrying water and is thus more efficient at this task than the military-derivative aircraft. Military-derivative aircraft were designed to carry cargo and troops, which are considerably less dense than retardant or water. Hence, military-derivative cargo aircraft typically hit their weight limit when carrying liquids well before using up all the volume inside the aircraft.

Although the acquisition cost of a 3,000-gallon commercial-derivative airtanker is similar to that of a 3,000-gallon military-derivative airtanker, the commercial derivative's O&S cost estimate is somewhat lower. This result is driven by our use of the Air Force's C-130J-30 as our proxy 3,000-gallon military airtanker. That aircraft has experienced somewhat higher maintenance costs per flying hour than have commercial aircraft of similar size.

Our 3,000-gallon fixed-wing military aircraft annualized cost estimate is higher than the Forest Service currently pays to contract for aging P2-Vs, as shown by the rightmost bar of Figure 4.1. The lower cost of the contracted aircraft may reflect the lower acquisition costs associated with these military surplus aircraft.

Helicopter Cost Estimates

Figure 4.2 presents the estimated annualized costs of three rotary-winged aircraft. Note that cost estimate for the 1,200-gallon helicopter (the leftmost bar in the figure) is similar to the fiscal year 2009 contract cost of a 1,300-gallon helicopter for the same number of days of service and hours of operation. The fiscal year 2009 contract costs of a 2,700-gallon helicopter are also shown with an escalation for assumed higher fuel costs of $4 per gallon. We were not independently able to estimate the costs of a 2,700-gallon helicopter because the available data sources did not provide enough information to develop such an estimate.

Our overall cost analysis considered retardant costs. It was important for us to consider these costs because one of our goals was to com-

Figure 4.2
Estimated Annualized Helicopter Costs

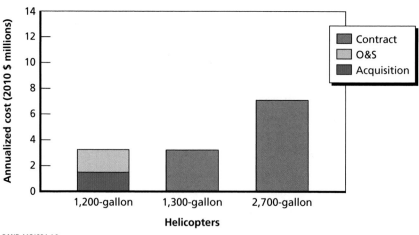

pare airtankers that drop retardant to helicopters and scoopers that do not. At $3 per gallon, these retardant costs can be considerable, as shown in Figure 4.3.

The cost of retardant ranges from about $1.50 to $3.00 per gallon. We used the higher value in our analysis because we believe it more accurately captures the full costs of retardant.[5] Scoopers and helicopters do not generally carry retardant. We received estimates that the foam often used in scoopers costs roughly $50 per 1,600-gallon load. We include foam costs in our optimization analyses in Chapters Five and Six, but they are considerably less important than the $9,000 per load of retardant in a 3,000-gallon airtanker.

Note that when retardant costs are included, the 1,500-gallon military-derivative airtanker is approximately 60 percent as expensive

Figure 4.3
Annualized Costs for Candidate Aircraft, Including Retardant Costs

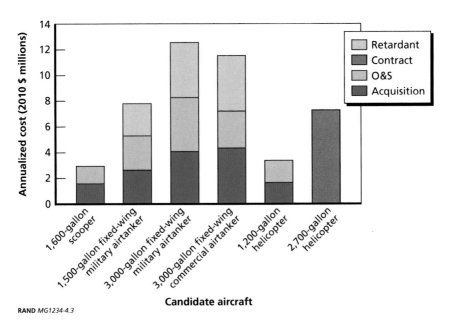

---

[5] Discussions with the Forest Service led us to believe that the high end of the range includes the cost of delivering the retardant to the operating locations and the aircraft, while the lower end includes only the costs of the retardant.

as the 3,000-gallon airtanker while offering half the carrying capacity. Everything else equal, the smaller military-derivative airtanker is therefore likely to be less cost-effective.

In the next two chapters, we present findings from the National Model and the Local Resources Model. Each chapter draws together its respective model results, the costs of large fires reported in Chapter Three, and the estimated costs of aircraft presented in this chapter to estimate an optimal initial attack fleet.

CHAPTER FIVE

The RAND National Model

As discussed in the Chapter One, we use two different but complementary approaches to estimate the social cost–minimizing portfolio of Forest Service initial attack aircraft. In this chapter, we present the RAND National Model, which uses Forest Service data to develop a fire simulation and then estimates the portfolio of aircraft that would minimize the total cost of wildfires, including cost of large fires (Chapter Three) and cost of aircraft (Chapter Four).

The next chapter, Chapter Six, presents the Local Resources Model. The Local Resources Model has a similar objective as the National Model, but it uses FPA's simulated fires. Unfortunately, as we discuss in that chapter, there are numerous practical challenges associated with using FPA in a manner for which it was not designed. Thus, the more parsimonious, purpose-built National Model discussed here has considerable advantages (e.g., in terms of its speed and ability to conduct "what-if?" excursions more easily). That said, there are realism and detail advantages to the Local Resources Model.

Both chapters have broadly similar methodologies. In both chapters, we estimate a relationship between available aircraft and the number of large fires, i.e., we characterize the function $f(a)$. We then use the large fire cost function presented in Chapter Three, $C_F(f(a))$, and the aircraft cost function from Chapter Four, $C_A(a)$, to search for the portfolio, a, that minimizes $C_F(f(a)) + C_A(a)$. The differences between Chapters Five and Six lie in the characterization of the function $f(a)$, how the number of large fires is modeled as a function of the aircraft portfolios, and, in $C_F(f)$, the relationship between the large fires that occur and their social costs.

Ultimately, both this chapter's findings and those reported in Chapter Six are based on assumptions and simplifications, as are any model's results. In undertaking these two complementary approaches, we were especially interested in commonalities and the insights that emanate from both models, despite their considerable methodological differences.

In this chapter, we begin by discussing the historical data that the National Model uses to simulate fires. Then, we discuss the estimation of fires' rates of spread, which we used to determine which fires should be attacked by aviation. Then, we describe how we estimated the total costs (of large fires and aviation) of different prospective portfolios of aircraft and how we undertook both national-level and geographically restricted estimations of the total cost-minimizing aviation portfolio.

Finally, we present the National Model's results. We start with an illustrative case that looks exclusively at airtankers. Then, we move to a portfolio analysis that examined the use of airtankers and scoopers and present both national-level and geographically restricted results. Geographically restricted results call for more aircraft because they make more conservative assumptions about how readily aircraft can relocate to fight fires.

Building a Fire Simulation

The National Model's fire simulation is based on historical fire data. We used data on wildfires in the United States in calendar years 1999–2008. The FPA Fire Ignition Generator (FIG) data set provides date, location, wind, and moisture levels associated with past fires. The fuel involved in each fire was assigned based on location using the 40 Scott and Burgan Fire Behavior Fuel Models layer of the LANDFIRE data system (see Scott and Burgan, 2005). We obtained information on slope, elevation, and distance to water sources for each of the historical fires using U.S. Geological Survey data on the wildfire ignition points listed in the FIG data set (see Fire Program Analysis System, 2009).

The National Model uses the FIRESTAT data set to simulate fires. We had, for instance, ten years of July data (1999–2008). We used those 310 days of fires to represent ten years of July fires, each with an observed location and estimated rate of spread. In essence, we did not so much simulate (in a Monte Carlo sense) these historical fires as *replay* them. Because we used the actual fires that occurred on actual dates, we incorporated Gilless and Fried's (1999) observation that there is intraday fire severity correlation, i.e., if one July 1 fire has a high rate of spread, it is more likely other July 1 fires do, too.

The National Model constructs ten synthetic fire years by extracting single-day samples from a ten-year historical data set of wildfires that occurred on federal and state lands. We individually tested each of these single-day fire scenarios against a prospective fleet of initial attack firefighting aircraft to estimate the fleet's efficacy. We then aggregated the resulting single-day tests to reflect a year's worth of fire costs to estimate the optimal mix.

In the National Model, each date is independent: A new simulated July 1 fire can be successfully attacked or not. Fires that are not contained on July 1 are considered to become large, and initial attack aircraft are no longer devoted to them. The outcome on July 2 is not affected by the outcome on July 1, because the model's process is assumed to "reset" daily. In our model, aircraft would not be tied up with "leftover" July 1 fires come the morning of July 2. Instead, July 2 is a new and independent date whose initial attack outcomes will not be affected by what preceded it. Similarly, Martell et al. (1984) assumed that their initial attack system began each day with all aircraft available for immediate dispatch.

Fires' Rates of Spread

Each simulated fire had a "real" outcome when it actually occurred (e.g., it became a costly large fire, or it was a small fire that was quickly contained, a much more likely scenario). The data available to us did not indicate which resources (ground or air) were used against these historical fires. Nor, of course, did we know the counterfactual of what would have happened if more or fewer resources were used against the fire.

We can, however, estimate each simulated fire's rate of spread, or how quickly it grew given the fuel, moisture, and wind conditions associated with it. We used Scott and Burgan's (2005) model to estimate individual fires' rates of spread, for which the Forest Service provided a spreadsheet implementation. Our examination suggests that this implementation provides rate-of-spread estimates that are similar to those provided by BehavePlus, a Forest Service–developed fire behavior model (see Andrews, 2009).

Scott and Burgan's (2005) approach assumes that the contents and topography of the fuel bed are both homogeneous and continuous and that weather conditions are unchanging. We assumed that the fires grew elliptically, with faster headfire rates of spread and slower flanking and backing rates of spread. We also assumed that the rate of spread did not change (accelerate or decelerate) over the simulated period. We did not have sufficient information to vary the fire growth rate by time of day.

We assumed that fires with a faster rate of spread are more likely to become large. In reality, some very fast-spreading fires were nevertheless contained, perhaps because of the extensive use of firefighting resources against them, or perhaps because of geography (e.g., the fire burned quickly to the edge of a river that it could not jump). Conversely, in reality, some fires may have had very slow rates of spread but burn undetected for long periods in isolated areas before bursting forth as large fires.

Unfortunately, the data available to us did not provide insight into the level of local resources that can fight any specific fire. (The Local Resources Model, by contrast, uses information about local resources embedded in FPA, as discussed in Chapter Six.) Instead, the National Model makes the simplifying assumption that baseline local firefighting resources are homogeneous across nascent fires. If one assumes that local firefighting resources are homogeneous, it logically follows that the fires with the fastest rate of spread will be those that become large, and this has been the case historically. But the National Model, with its homogeneous baseline resource assumption, predicts that no low-rate-of-spread fires will become large, a prediction not borne out by the data.

Nevertheless, the National Model assumes that a fixed, baseline level of local fire-line production capability is available against any fire. These local resources include on-the-ground firefighters with associated equipment, smaller Type II and Type III helicopters, and state and local aviation assets. According to the model, the majority of fires would be contained by these local resources.

Large aircraft can supplement these local resources to increase the rate at which a fire line is built. When a fire line is built more quickly, the fire is more likely to be contained.

Our analysis did not compare the relative desirability of incremental investments in local firefighting resources (e.g., firefighters, fire trucks) versus aviation. Instead, we held the level of local resources constant and assessed the incremental value of aviation. The much more granular California Fire Economics Simulator described by Fried, Gilless, and Spero (2006) and Haight and Fried (2007) provides insights into issues such as the deployment of fire engines, dispatch rules, and line-building tactics.

Albini, Korovin, and Gorovaya (1978) presented an approach to determining the time required to contain a fire when the fire's rate of spread and the fire line production rate are specified. FPA's Initial Response Simulation uses the approach outlined in Fried and Fried (1996), which is an extension of Albini, Korovin, and Gorovaya's. We used Albini, Korovin, and Gorovaya's approach because we did not have sufficient data on each fire to fully implement the Fried and Fried model.

Intuitively, fire containment is a race. The fire grows at a specified rate while firefighters attempt to build a fire-control line to contain it. If enough fire line can be built sufficiently quickly, the initial attack succeeds and the fire is contained. Otherwise, the fire becomes large.

Airtankers, scoopers, and helicopters can help "win these races" by supporting the fire-line production with their retardant or water drops. In accord with FPA, the National Model assumes that every 100 gallons of retardant aids in the production of one additional chain of fire line on grass.[1] The efficacy of aerial drops varies with the composition of the fuel onto which the drop occurs (e.g., drops onto grass are more effective than drops onto trees). We accounted for this phenomenon by reducing fire-line production based on fuel composition in accord with FPA's approach.[2] Further, we assumed that water is less effective than retardant in producing fire line. Therefore, we depreciated the fire-line production rate for water-dropping scoopers and helicopters by 50 percent in accordance with the approach taken by Fire Program Solutions (2005). Later in this chapter, we present the results from our sensitivity tests on this parameter. As we discuss in Chapter Six, FPA does not depreciate water efficacy relative to retardant efficacy.

Sending Aircraft Against Fires

Suppose that the Forest Service has a portfolio of firefighting aircraft. On each day of a fire season, dispatchers must decide which aircraft to send against which fires. Such decisions are fraught with uncertainty (e.g., there is likely to be considerable uncertainty as to which nascent fires are most dangerous). Indeed, a fire may be too fast-moving, and even if firefighting aircraft were sent against it, "the race" cannot be won. Alternatively, a fire may grow slowly enough that local resources would be enough to contain it without the use of large aircraft. Ideally, firefighting aircraft would be sent against those nascent fires where they can make the difference between escape and containment, rather than fighting fires that would escape or be contained irrespective of aviation usage. However, perfect foresight as to which fires would most benefit from firefighting aircraft is unlikely.

---

[1]   According to National Wildfire Coordinating Group (2011), one chain equals 66 feet, or 20 meters, so 80 chains equals one mile of fire control line.

[2]   According to FPA's production factors, 100 gallons of retardant creates one chain of fire line on short grass, perennial grass, or western woody shrub. The same amount of retardant creates 0.6 chains of fire line on grass with pine, sawgrass, tundra, or pine litter or 0.4 chains of fire line on all other fuels, including hardwoods, pine, and slash. See Fire Program Analysis System (1995).

We categorized new fires by their relative need for air support. In our vernacular, a Category A fire is a fire that will be contained by local resources, though large aircraft can make containment come sooner. A Category B fire, by contrast, will become large if only baseline local fire-line production resources are used, but large aircraft can augment those resources to achieve containment. A Category C fire will become large irrespective of large aircraft use. Category C fires have the fastest rates of spread; Category A fires have the slowest. The appellations Categories A–C are assigned to new, small fires. The model assumes that Category A fires will remain small and Category C fires will become large. Category B fires are, in some sense, the most interesting fires because it is on these fires that aircraft usage can be the difference between the fire being contained while small or becoming large and potentially very destructive.

In reality, the fire dispatch system will not know a new fire's category. Although one might want to send aircraft only to Category B fires, where they make the most difference, we assume in this chapter that aircraft, if they are available, are sent to all Category B fires, as well as to all Category C fires (that nevertheless become large) and "close-call" Category A fires (where aircraft are not needed to achieve containment, but it is close, i.e., within 10 percent of the rate of spread for which aviation would be required). In the scenarios, the close-call Category A fires receive one drop. These drops are, in some sense, wasted, because Category A fires do not need drops to be contained. Category C fires receive the number of drops equivalent to the capability of one aircraft for one day. (We do not, however, assume that the aircraft would continue to fight a now-large fire after its first day.) As with the Category A drops, drops on Category C fires are wasted in our model, but we believe that they are important to include to make our dispatch process more realistic. Figure 5.1 shows how the National Model impedes aircraft dispatch prescience and offers a visual comparison between prescient and realistic dispatch capabilities.

If there are more fires that request aircraft than there are aircraft available, the aircraft are assigned to fires at random (i.e., Category B fires do not receive their aircraft with higher priority than other fires in the other categories).

Figure 5.1
Prescient Versus Realistic Aircraft Dispatch in the National Model

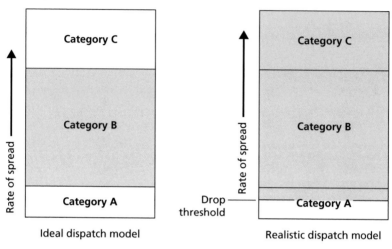

RAND MG1234-5.1

Table 5.1 shows a mixed verdict on the empirical validity of our categorization scheme. In theory, no Category A fires should become large. In fact, 3.3 percent of our "regular" Category A fires became large, and 4.7 percent of our close-call Category A fires became large. Our model also predicts that all Category C fires will become large. In fact, about 11.4 percent became large. One explanation for the gap between theory and reality is the assumption of homogeneous local resources. Many Category C fires were contained because they received more firefight-

Table 5.1
Theoretical Versus Actual Large Fires

Category	% of Fires	Theoretical % Large Fires	Actual % Large Fires
Regular A	87.0	0	3.3
Close-call A	1.4	0	4.7
B	10.0	Varies with level of aviation resources	6.0
C	1.5	100	11.4

ing resources than we assumed in our model, and the converse was true for Category A fires that became large. It is, however, encouraging that the results in the rightmost column in Table 5.1 are in ascending order, meaning that fires with a higher rate of spread have been more likely to become large.

Table 5.1 also shows that the vast majority of fires are in Category A, the lowest rate-of-spread category. Most of these fires would not become large even if there were no Forest Service aviation brought to bear against them.

We assumed that the baseline local-level rate of fire-line production is a calibration parameter in the National Model. We calibrated the model to reproduce historical large fire frequencies. We know that some fires will become large irrespective of aircraft usage, however. Gebert, Calkin, and Yoder (2007) reported on 3,061 large wildland fires between 1995 and 2004. Prior to 2004, the Forest Service had access to more than 40 3,000-gallon airtankers, so most requests for airtanker support were fulfilled. The Forest Service provided data indicating that, prior to the 2002 crashes, the rate at which airtankers were unable to fulfill requests was about 7 percent. With today's smaller fleet, this rate now averages 22 percent. Therefore, we assumed that roughly 300 large fires per year would occur even with 40 or more 3,000-gallon airtankers.

We varied the assumed local baseline rate of the fire-line production parameter to achieve the 300 large fires per year asymptote when airtankers are abundant. This parameter value ended up being 60 chains of fire line per hour—that is, in the absence of large aircraft, we assumed that local resources generated an average of 60 chains of fire-control line per hour. We defined the close-call Category A cutoff as 10 percent lower where rates of fire spread are equivalent to 54 chains per hour.

Testing a Portfolio of Aircraft

Our objective was to estimate the Forest Service's cost-minimizing portfolio of aircraft, or how many and what kind of aircraft would minimize the total social costs of wildfires, including the cost of large fires and the cost of large aircraft. (Chapters Three explained how we estimated the costs of large fires, and Chapter Four provided an overview of the representative large aircraft and their cost estimates. In this chapter and the next, we use annualized cost estimates for new airtankers and scoopers and leased helicopters.)

Figure 5.2 shows how we used our fire simulations to evaluate the different prospective aircraft portfolios.[3] Any prospective portfolio of aircraft has costs and also affects the number of large fires. Total costs that we attempted to minimize are the sum of aircraft costs and large fire costs.

The National Model starts with a prospective portfolio of firefighting aircraft. A year's worth of fires is then simulated based on historical fire data. Each day in the simulated year has a certain number of new fires (with the number of new fires peaking in the summer months, in accordance with reality). The prospective portfolio's firefighting aircraft are dispatched against these fires. Helicopters and

Figure 5.2
Structure of Initial Attack Simulation

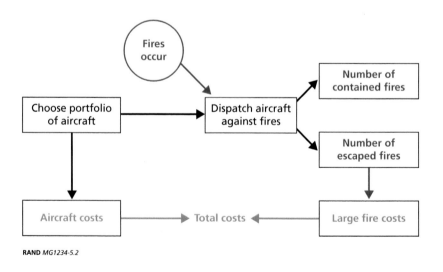

RAND MG1234-5.2

---

[3] The structure shown in Figure 5.2 is equally valid for the Local Resources Model, described in Chapter Six.

scoopers, unlike airtankers, require water proximity, so available helicopters and scoopers are sent against fires where they would be effective, with the more versatile airtankers dispatched to remaining fires that may or may not be water-proximate. It is possible that all of a given day's simulated new fires can be contained by the portfolio of firefighting aircraft. Alternatively, the portfolio could run out of aircraft, and some fires become large and costly. In both the National Model and the Local Resources Model, a fire is said to become large if it cannot be contained in one day.

Kliment et al. (2010) found that Forest Service airtanker mission times have averaged 45 minutes. We assumed that a single airtanker can perform no more than seven missions in one day. We selected this parameter based on the 45-minute average mission duration and an understanding that airtankers operate for only eight hours a day. Seven missions do not account for all of an airtanker's time because the balance of the day is spent on activities such as refueling.

We did not want our aircraft to be unrealistically effective and therefore built in slack. For example, we imposed the seven-missions-per-day constraint to account for real-world imperfections, such as inaccurate drops. Also, as discussed earlier, all of our aircraft "waste" drops on close-call Category A and hopeless Category C fires.

For helicopters and scoopers, the production rate calculation is more complicated because these aircraft cycle between a water source and a fire, not between a base and a fire. We calculated the helicopter- or scooper-generated fire-line production rate by determining the cycle time between the fire and the nearest body of water.

For helicopters, we assumed that the total fixed cycle time is six minutes and that a helicopter travels at an average of 80 miles per hour, a value between the laden and maximum speed. In the National Model, we did not vary aircraft performance parameters (e.g., how much a helicopter could carry or how fast it could fly) with elevation or temperature. Our aircraft performance parameters were representative of typical operating conditions. We did not perform detailed calculations on the impact of atmospheric conditions on aircraft performance. We assumed that scoopers have a fixed cycle time of five minutes and an average air speed of 150 miles per hour.

Figure 5.3 shows the number of gallons per hour that our model estimates could be dropped by a 2,700-gallon helicopter, a 1,600-gallon scooper, and a 3,000-gallon airtanker as a function of distance to the nearest water source subsequent to arrival at the fire. Of course, an airtanker is unaffected by distance to water. It will, on average, drop 4,000 gallons of retardant per hour, assuming a 45-minute cycle time. Suppose a fire is located ten miles away from the nearest scooper- or helicopter-accessible body of water. In this case, after each aircraft's first drop, helicopters would cycle every 21 minutes (six minutes fixed plus 15 minutes in flight—7.5 minutes each way covering ten miles), while scoopers would cycle every 13 minutes (five minutes fixed plus eight minutes in flight—four minutes each way covering ten miles). With a 21-minute cycle time, a 2,700-gallon helicopter would drop about 7,700 gallons per hour, while, with a 13-minute cycle time, a 1,600-gallon scooper would drop about 7,400 gallons per hour. (Note, however, that more bodies of water are accessible to helicopters than to scoopers, so it could be that helicopters would have a greater advantage than this illustrative discussion suggests.)

Figure 5.3
Gallons Dropped per Hour After Arrival at Fire Site, by Type of Aircraft

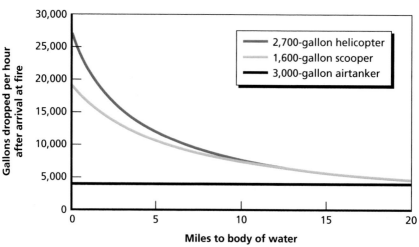

Figure 5.3 does not account for the time required to arrive at the fire site. We assumed that airtankers could reach and attack a new fire one hour after detection. Given their slower air speeds, we assumed that scoopers could reach a new fire one and a half hours after detection, while helicopters could reach a new fire two hours after detection.

At the end of the simulated fire season, the model tallies the number of large fires and the costs of the prospective portfolio of aircraft. These statistics trade off against one another: More aircraft cost more but imply fewer large fires.

Different prospective portfolios of aircraft (e.g., different numbers and types of aircraft) can then be tested to identify the portfolio of aircraft that minimizes the total social costs of wildfires.

Base Version Versus Restricted Variant of the National Model

Our base version of the National Model assumes that any aircraft can fight any fire. On some level, this assumption is a truism. The real issue is how quickly the aircraft could get to a fire. As noted, we parameterized the National Model using Kliment et al.'s (2010) finding that airtanker mission times have averaged 45 minutes. Of course, if an airtanker had to fly from, say, western Montana to southern Arizona, this assumption would not hold. The base version of the National Model therefore overrates the capabilities of airtankers to quickly deploy anywhere in the United States where they might be needed. This shortcoming would be especially acute with a small fleet of airtankers. The observed 45-minute average mission time comes from current airtanker fleet sizes, which allow airtankers to be spread across the country in such a way that lengthy missions are generally not necessary. The base version of the National Model therefore provides more sensible results with fleet sizes akin to the current size than with considerably smaller fleets. For similar reasons, the base version of the National Model overrates the value of individual scoopers and helicopters.

To address this concern, we created a variant of the National Model that restricts individual aircraft to operating within a single

Forest Service Geographic Area Coordination Center (GACC) in a
given month. Figure 5.4 is a map of Forest Service GACCs. Each dot
represents the location of a geographic area's headquarters.

The Eastern and Southern GACCs are quite large geographically
but do not concern us greatly. The Southern GACC's fire season tends
to be in the spring, when other GACCs have relatively few fires; the
Forest Service's airtanker fleet can be located in the Southern GACC
in the spring to handle fires there. Much of the Eastern GACC, except
northern Minnesota, has relatively few fires that call for Forest Service
involvement.

Each of the nine western GACCs is sufficiently geographically
compact that the assumption of a 45-minute average airtanker mission
time is more tenable. Certainly, a modern fixed-wing aircraft could fly

Figure 5.4
Forest Service GACCs

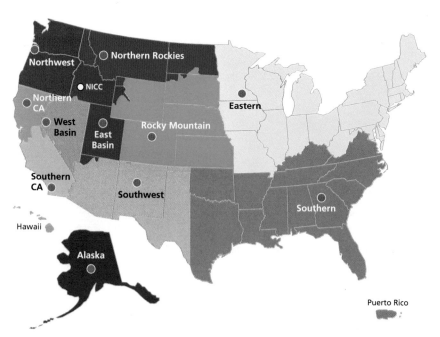

SOURCE: NIFC, undated(b).
NOTE: NICC = National Interagency Coordination Center.
RAND MG1234-5.4

between most points within any single western GACC in less than one hour.

Indeed, restricting aircraft to a single GACC in a given month is probably an overly onerous parameter. Suppose, for instance, that the East Basin GACC was quiet but the West Basin GACC had a considerable number of new fires. It would be eminently reasonable for an airtanker based in Salt Lake City to help fight a fire in eastern Nevada, but the GACC-restricted variant of the National Model would not allow such a reasonable step.

In short, the base version of the National Model overrates the value of individual aircraft and might therefore suggest an unrealistically small fleet size, and the National Model GACC-restricted variant has exactly the opposite bias. We believe that the two approaches' total cost-minimizing solutions bracket the truth, conditional on the validity of the other assumptions undergirding the National Model.

National Model 3,000-Gallon Airtanker Total Cost Minimization Illustration

To illustrate how the National Model's total cost minimization works, we use the base version's 3,000-gallon airtanker case. The parameters, or assumptions, of this exploration are as follows:

- Geography is ignored.
- There are 3,000 gallons of retardant per drop.
- Aircraft attack Category C fires, Category B fires, and close-call Category A fires.
- Acquisition and O&S costs per aircraft run $7.1 million per year.
- Retardant costs $3 per gallon.
- The average large fire cost is $3.3 million.

As shown in Figure 5.5, more 3,000-gallon airtankers imply fewer large fires. With a large number of 3,000-gallon airtankers, the National Model is calibrated so that there would be approximately 300 large fires per year.

Figure 5.5
**Estimated Relationship Between 3,000-Gallon Airtankers and Average
Annual Large Fires**

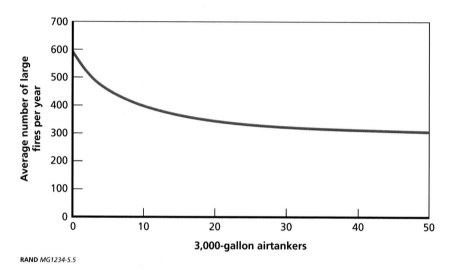

RAND *MG1234-5.5*

The first few 3,000-gallon airtankers have a considerable effect, sharply reducing the number of large fires. But additional aircraft have a lower incremental effect, and the asymptote at 300 large fires is achieved. The National Model overrates the impact of the first few airtankers because it does not consider the increased flight times that would doubtlessly accompany such a small fleet of aircraft. As the fleet size grows, a 45-minute average mission time becomes a more tenable assumption, and results are therefore more valid.

This exploration assumes that 3,000-gallon airtankers are the only large aircraft involved with firefighting. Furthermore, this exploration considers only the initial attack phase of small fires; it does not address aircraft usage in extended, multiday campaigns against large fires.

As shown in Figure 5.6, this National Model exploration indicates that expected total costs would be minimized with 23 3,000-gallon airtankers used in initial attack.

Figure 5.6 plots annualized aircraft costs (the rising green line), retardant costs (the orange curve), and estimated annual large fire costs

Figure 5.6
Estimated Cost-Minimizing Number of 3,000-Gallon Airtankers Devoted to Initial Attack

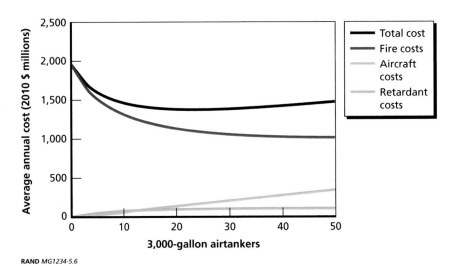

RAND *MG1234-5.6*

(the red curve), with the black curve representing total costs, summing the three. The black curve's minimum is at 23 aircraft.

Note that this tally is only for 3,000-gallon airtanker aircraft used in initial attack. Next, we consider airtankers in conjunction with helicopters and scoopers.

Helicopters and Scoopers in the National Model

Airtankers are not the only firefighting aviation option. In this section, we additionally consider the use of helicopters and scoopers in initial attack.

Helicopters and scoopers cannot completely fill the Forest Service's initial attack requirements. Some fires are too far from water for scoopers and helicopters to be effective. The National Model's algorithm sends scoopers and helicopters to fires where they are effective, saving airtankers for other fires. The issue, then, is the total social cost–minimizing combination of airtankers, scoopers, and helicopters.

Traditionally, helicopters have focused on extended campaigns against large fires, a scenario outside our modeling approach. In that sense, the National Model misses the predominant portion of what helicopters do best. We also ignore ancillary missions that helicopters can perform (e.g., moving equipment and personnel).

After comparing historical fire locations in FIRESTAT with the sizes and locations of water bodies from the U.S. Geological Survey data (see Esri, 2012), we found that at least two-thirds of historical fires have been within ten miles of a scooper-accessible body of water, and about 80 percent have been within five miles of a helicopter-accessible body of water. These water-proximate fires are those against which helicopters and scoopers would be most valuable, as shown in Figure 5.3. Figure 5.7 shows the proximity of the fires in our historical data set to bodies of water meeting aircraft requirements, including lakes, rivers, and oceans (e.g., the Gulf of Mexico, the Pacific Ocean). We included seasonal water bodies without regard to season and have made no attempt to adjust the suitability of a water source based on its likely size at a given time of year. Figure 5.7 also assumes that the

Figure 5.7
Historical Fires' Proximity to Bodies of Water of Varying Sizes

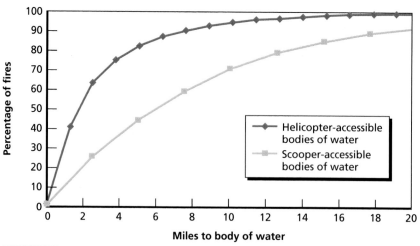

Forest Service would have permission to draw from these bodies of water. Figure 5.7 may therefore be optimistic about water availability.

According to our interviews, 12 feet is considered to be the minimum diameter of a body of water that can be accessed by a helicopter. A 1,600-gallon scooper requires a length of 4,400 feet to descend empty, retrieve a full load, and ascend to a height of 50 feet so as to remain clear of potential obstructions surrounding the body of water (Gonsalves, 2009).

In our simulations, we found that scoopers could draw from freshwater lakes 81.2 percent of the time, from rivers 16.4 percent of the time, and from saltwater sources (most frequently the ocean but also, potentially, the Great Salt Lake) about 2.4 percent of the time. Likewise, we found that helicopters draw from freshwater lakes 92.2 percent of the time, from rivers 6.6 percent of the time, and from saltwater sources 1.2 percent of the time.

Further abetting the argument for scoopers, most human settlement is near water. Therefore, the most worrisome fires in terms of the threat to large numbers of houses tend to be water-proximate. Illustrating this phenomenon, Figure 5.8 depicts the region around Phoenix, Arizona. The red dots indicate historical fire locations, while the blue areas are scooper-accessible water sources. Even in arid Arizona, scooper-accessible water sources are available near this large city.

It is generally only in remote, sparsely populated areas where water availability would be problematic that airtankers would be required.

Figure 5.9 presents the National Model's total cost-minimizing combination of 3,000-gallon airtankers and 1,600-gallon scoopers. The model reveals that a fleet of five 3,000-gallon airtankers and 43 1,600-gallon scoopers optimally minimizes the total social costs of wildfires.

The 3,000-gallon airtankers–only curve in Figure 5.9 is the same 3,000-gallon airtanker total cost curve shown in Figure 5.6. There are considerable cost savings in being able to use scoopers rather than relying solely on airtankers.

Figure 5.9 suggests that the Forest Service's most cost-effective fleet would be composed largely of scoopers with a fairly small number

Figure 5.8
Scooper-Accessible Water Sources and Historical Fire Locations Near Phoenix, Arizona

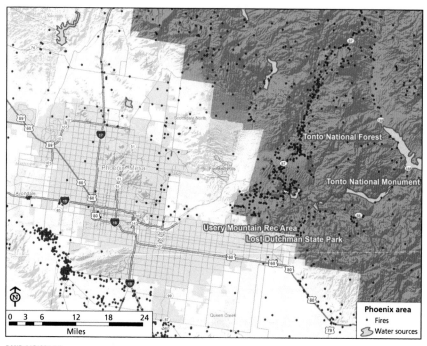

of airtankers in an ancillary role to fight fires not proximate to scooper-accessible water sources.

We also ran an analog to Figure 5.9 in which we allowed for a portfolio of airtankers and 2,700-gallon helicopters. The resultant portfolio of 16 airtankers and 15 2,700-gallon helicopters was more costly than the five-airtanker, 43-scooper optimum in Figure 5.9.

We were not able to run three-aircraft searches with the National Model—airtanker, scooper, and helicopter—because the computational complexity of the model is multiplicative with the number of types of resources (e.g., the number of different types of aircraft examined). Given that airtankers and scoopers were found to be more cost-effective than airtankers and helicopters, we focus on airtankers and scoopers in the remainder of this chapter.

Figure 5.9
**Total Cost-Minimizing Combination of 3,000-Gallon Airtankers and
1,600-Gallon Scoopers in the National Model**

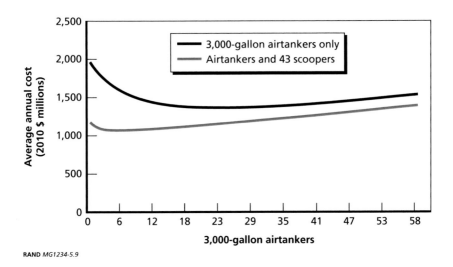

RAND *MG1234-5.9*

Table 5.2
**Result Sensitivity to Different Large Fire Average Costs in the National
Model**

Average Cost of a Large Fire	3,000-Gallon Airtankers	1,600-Gallon Scoopers	Total Social Cost (2010 $ millions)
$0.3 million	0	8	154
$2.1 million	4	36	739
$3.3 million	5	43	1,073
$4.5 million	6	55	1,935
$10 million	8	57	2,830

NOTE: The shaded row represents the base-case estimate.

We were able to assess how optimal portfolio choices vary with the average cost of a large fire. Table 5.2 contrasts the chosen number of aircraft associated with average large fire costs of $2.1 million, $3.3 million, and $4.5 million, our range of large fire cost estimates discussed in Chapter Three, along with very low and high values of $300,000 and $10 million.

In the $2.1 million to $4.5 million range, the portfolios are similar, with a relative emphasis on scoopers and airtankers in an ancillary role.

The National Model's Sensitivity to Assumptions of Water Efficacy

As discussed earlier, there is a perceived difference in the per-gallon efficacy of scoopers and helicopters when compared with airtankers because airtankers drop retardant instead of water. There was a consensus view in our interviews with Forest Service personnel and others that retardant is twice as effective as water in fire-chain production, though there was some discussion that water is even less effective in some environments. However, the FPA model used in the Local Resources Model, discussed in Chapter Six, specifies a one-to-one efficacy ratio for retardant-to-water. Thus, it is very important to understand the sensitivity of the model results to this assumption. Table 5.3 compares the National Model results for different assumptions about the relative efficacy of water.[4]

Table 5.3 indicates that there is a strong preference for a scooper-centric fleet when water is considered to be at least one-quarter as effective as retardant. If water is considered to be less than 20 percent as effective as retardant, the optimum moves toward Figure 5.6's airtanker-centered fleet (though, even at 5-percent efficacy, a sizable number of scoopers would be requested). Because there was a general consensus that water is about half as effective as retardant, there should be a preference for a scooper-centric fleet, according to the National Model.

---

[4] We did not vary retardant efficacy. Instead, we used the FPA-employed retardant efficacy levels discussed earlier in this chapter. The uncertainty evaluated in Table 5.3 is how water efficacy varies as a percentage of the assumed level of retardant efficacy.

Table 5.3
Sensitivity to Assumptions of Water and Retardant Efficacy in the National Model

% Efficacy of Water Compared to Retardant	3,000-Gallon Airtankers	1,600-Gallon Scoopers	Total Social Cost (2010 $ millions)
100	2	40	594
75	2	40	811
50	5	43	1,073
33	7	44	1,171
25	9	43	1,226
20	13	37	1,249
5	18	30	1,298
0	23	0	1,371

NOTE: The shaded row represents the baseline efficacy assumption in the model.

An interesting characteristic of Table 5.3 is that more scoopers are demanded when water is 33 or 50 percent as effective as retardant than when it is 75 or 100 percent as effective. In that range of values, the effect of reducing water efficacy is to increase the demand for scoopers (because more water is needed to achieve desired outcomes). Thus, it is only when water becomes highly ineffective that there is a net decrease in demand for scoopers.

Table 5.3 also provides insight into the possible effect of making fewer water sources available to scoopers. Suppose, for instance, that N percent of the water sources we have identified as being available to scoopers are not, in fact, available to them (e.g., they are privately owned and permission cannot be obtained to draw from them). For $100 - N$ percent of fires, there would be no effect; their chosen water source could be utilized. For N percent of fires, a new water source would have to be chosen. However, that alternative water source could be quite close to the chosen water source, in which case the effect would simply be a modest increase in scooper cycle times and a modest decrease in water efficacy. Table 5.3 suggests that the effect of such a water efficacy decrease could be an increase, not a decrease, in the number of scoopers demanded. For large values of N in which many fires need different water sources and those alternative water sources could be a considerable distance away, one might see a diminution in demand for scoopers.

A complementary way to think about water availability would be if the average scooper had to fly twice as far as we are assuming (i.e., if we have grossly overestimated the number of scooper-accessible bodies of water). Such a scenario would correspond to halving water efficacy in Table 5.3. But going from our baseline, 50-percent efficacy to 25-percent efficacy would cause no change in the number of scoopers required (43). We conclude that our recommendation for a scooper-centric initial attack fleet is robust to a considerable diminution in water availability.

Figure 5.10 illustrates how the cost-effectiveness of a mixed fleet that uses scoopers is sensitive to the effectiveness of water relative to retardant. The gap between a mixed fleet and an airtanker-only fleet narrows substantially if water is thought to be less than 30 percent as effective as retardant.

Figure 5.10
Mixed Fleet Sensitivity to Water Efficacy Assumptions in the National Model

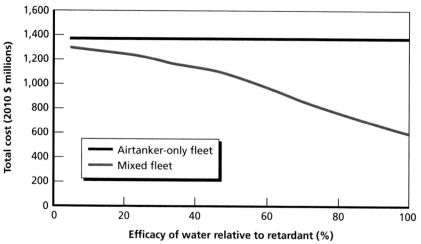

Aircraft Allocation in the National Model Restricted Variant

Returning to our baseline two-to-one retardant-to-water efficacy ratio, the National Model GACC-restricted variant found an overall cost-minimizing solution of eight 3,000-gallon airtankers and 48 1,600-gallon scoopers, versus five 3,000-gallon airtankers and 43 1,600-gallon scoopers in the base version of the National Model, as shown in Figure 5.10. In the GACC-restricted variant, we have different geographic assignments of the eight 3,000-gallon airtankers and 48 1,600-gallon scoopers for each month. Figure 5.11 shows how we propose distributing the 48 1,600-gallon scoopers in July.

We estimated a somewhat different allocation of the aircraft for August with, for instance, three of the scoopers moving with the fire

Figure 5.11
Proposed July Distribution of 48 1,600-Gallon Scoopers, by GACC

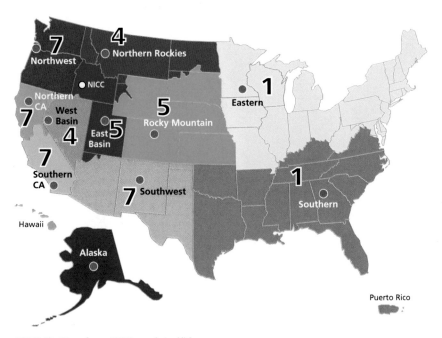

SOURCE: Map from NIFC, undated(b).
NOTE: NICC = National Interagency Coordination Center.
RAND MG1234-5.11

season from the Southwest GACC to the Northern Rockies GACC. Likewise, the model allocates the eight airtankers by month.

July and August are the most important months for aircraft location decisions because aircraft utilization is highest in these months. Aircraft availability constraints are less likely to bind in other months, when fire demands are generally lower.

Boychuk and Martell (1988) found a similar result when analyzing the hiring of firefighters in Ontario. When hiring can be provincially centralized, the total cost and optimal hiring levels are lower than when firefighters are restricted to operating within a specific region of the province. The National Model GACC-restricted variant parallels their region-restricted results.

Table 5.4 shows the National Model GACC-restricted variant's optimal fleets for different average large fire costs. It favors a somewhat smaller fleet at $2.1 million per large fire, the low estimate in Table 3.3 in Chapter Three. The estimated optimal fleet is nearly the same, at $3.3 million per large fire (our baseline estimate) and at $4.5 million per large fire (the high estimate in Table 3.3).

There are at least two concerns with these National Model results. First, even the GACC-restricted variant does not consider aircraft basing issues. Second, the National Model does not consider differential firefighting resources at different locations. We developed the Local Resources Model to address these concerns.

Table 5.4
Cost-Minimizing Fleet Sizes with Varying Average Cost of a Large Fire in the GACC-Restricted Variant of the National Model

Average Cost of a Large Fire	3,000-Gallon Airtankers	1,600-Gallon Scoopers
$2.1 million	4	43
$3.3 million	8	48
$4.5 million	9	48
National-level $3.3 million	5	43

CHAPTER SIX

The RAND Local Resources Model

Like the National Model, the Local Resources Model systematically analyzes different aircraft combinations and basing options, identifying the fleet mix that minimizes total social costs, $C_F(f(a)) + C_A(a)$. However, unlike the National Model, the Local Resources Model's characterization of $f(a)$ is built on fire season, ground resources, and containment estimates generated by the FPA system. In addition, the Local Resources Model allows the differential social costs of fires, $C_F(f)$, to vary by location.

FPA is a Forest Service system designed to assist decisionmakers in resource allocation choices. FPA's traditional focus has been at the local Forest Service Fire Planning Unit level. From a local decisionmaker's perspective, nationally managed large aircraft availability is exogenous. Currently, FPA's simulations assume that there is a 60-percent probability that a request for large aircraft will be fulfilled. Local decisionmakers do not decide how many large aircraft the Forest Service operates, so FPA was not designed to consider such issues.

Our use of FPA in the National Resources Model is a considerable departure from its intended use in that our central interest is the value and number of large aircraft the Forest Service should have. To accommodate our interest, we overrode FPA's 60-percent large aircraft request fulfillment assumption. Instead, on a simulated fire-by-fire basis, we asked FPA to tell us how initial attack outcomes (large or small) differ under a range of conditions: without any large aircraft, with a 3,000-gallon airtanker available, with one or two 1,600-gallon scoopers available, or with a 2,700-gallon helicopter available. Logically, having a large aircraft available will never worsen a fire outcome; if a fire is contained without large aircraft, it will be contained with it. The most interesting cases are those in which large aircraft changed the expected outcomes (e.g., in which the fire went from large fire to small). There are also fires from which escape is inevitable; that is, even when fought with large aircraft, the fire becomes large.

As discussed in Chapter Five, we use the vernacular "Category A" to refer to fires that would be contained even without large aircraft, "Category B" to refer to fires against which large aircraft make the difference between a small fire and a large fire, and "Category C" fires for those that will become large irrespective of the use of large aircraft.

A key attribute of FPA is that it includes fire-by-fire estimates of local firefighting resources. These local resource estimates include the number of on-the-ground firefighters and their equipment, as well as the local availability of smaller aircraft (e.g., Type II and Type III helicopters). Smaller aircraft are managed locally, not nationally, so FPA considers their presence or absence for each simulated fire. FPA uses estimates of current inventories of local firefighting assets, their associated dispatch locations, and their seasonal availabilities.

The Local Resources Model is a set of algorithms that we developed to work in conjunction with the FPA model. The Local Resources Model is built on, and dependent upon, the FPA model and its data and methods for assessing the incremental contribution of different types of large aircraft against specific fires. We add routines for tracking and dispatching a given fleet of Forest Service aircraft to the FPA model. By establishing the distances that aircraft will need to travel to their bases, the Local Resources Model algorithms provide FPA information on the hourly rate at which water or retardant will be delivered by Forest Service air assets. However, it is FPA's algorithms, not the Local Resources Model's, that ultimately estimate whether those aviation-provided drops change fire size outcomes. For example, FPA assumes that one gallon of retardant has the same ability to build a fire-control line as one gallon of water. Similarly, FPA makes no distinctions between direct and indirect attack, nor does it allow for the possibility of aircraft slowing or suppressing a fire before ground crews can arrive. We could not alter these assumptions, which are intrinsic to FPA.

The Local Resources Model is a much more computationally intensive model than the National Model. It relies on a combination of servers and databases to conduct an analysis of even a single scenario. Some of those servers included FPA's "live" production servers. Because the Local Resources Model relies on the production version of FPA, we had limited flexibility to adapt the model to test the sensitivity of such parameters as the efficacy of retardant versus water or the method of attack, nor could we increase the number of scenarios examined.

Fundamentally, the National Model and the Local Resources Model rely on the same ten-year historical set of federal and state wildland fires. However, the Local Resources Model constructs synthetic fire years by sampling historic fires by location with ±3 days of sampling flexibility around the target date.[1] Using an eight-kilometer geographic grid, the Local Resources Model then conducts a random draw on a particular fire within each grid to select a fire that will determine the weather (wind and fuel moisture levels) for fires within that cell. Finally, the Local Resources Model conducts random draws on each fire to determine whether it will occur or not (based on statistical cause histories), how many fires occurred on a day, and the time of ignition. For production purposes, FPA creates 200 scenarios representing 200 synthetic years of fires. For technical reasons, our access to FPA was limited to the first five scenarios (years).

For each fire-ignition scenario, we prepared a menu of air assets that we then tested against each scenario in FPA's Initial Response Simulator (FPA-IRS). The menu of assets included no air support, one tanker, one helicopter, and one or two scoopers. Finally, we analyzed each combination of costs and benefits using an integer program, which ultimately identified the mix of assets and locations that resulted in an optimal investment in aircraft for initial attack.

The Local Resources Model uses two core modules of FPA. The first, the Fire Event Simulator, synthesizes fire seasons and fire starts

---

[1] This paragraph draws on information from the Fire Program Analysis System web site (2011).

based on historical information. It conducts random draws of historical fires and an associated set of historical weather conditions. Fire behavior is modeled using local information on fuel types, burn indexes, and estimates of corresponding fire rates of spread. The Local Resources Model does not change the simulator's results; it takes them as given.

The second core model is FPA-IRS, which simulates the effect of available firefighting resources on fire outcomes. It produces estimates of the total size of the fire, along with the amount of time that resources are assigned to contain it. It uses local dispatch locations of firefighting resources and their planned availabilities to attack fires. It then augments these local resources with requested nationally managed assets (e.g., large aircraft, smoke jumpers) to further improve wildfire containment efficacy. The key trade-off simulated in FPA-IRS is the size of the fire perimeter versus the timing and quantity of the fire-control line produced to contain the fire. We manipulated the types of large aircraft available to fight a simulated fire and the rate at which those aircraft deliver water or retardant.

Another key advantage of the Local Resources Model relative to the National Model is that it considers different values at risk (i.e., what would be destroyed by a large fire) by fire location. In the Local Resources Model, if a nascent fire were to become large, we would assign it a total social cost scaled to be proportional to its Stratified Cost Index (SCI) value, which is calculated from fire characteristics such as fuel type, elevation, slope, and, importantly, the value of housing within 20 miles of the fire's location. Gebert, Calkin, and Yoder (2007) of the Forest Service developed the SCI. A fire's characteristics enter a regression equation that yields an SCI value for that fire (if it were to become large). We might think of these fire-specific SCI values as point totals, with a larger value suggesting a potentially more costly large fire.

Use of the SCI allows a more subtle description of the function $C_F(f(a))$. Whereas the National Model approach described in Chapter Five simply assumed this that function was linear in terms of the number of large fires, here we are able to assign different costs to different large fires based on where they occur.

We attach social cost estimates to individual fires in a way that is proportional to SCI values, scaling those estimates so that the average cost of a large fire not contained by aircraft in the FPA simulation is $3.3 million, the average of the high and low estimates of the average cost of a large fire in Table 3.3 in Chapter Three. Using the empirical distribution of large fire costs identified by Gebert, Calkin, and Yoder (2007), as well as subsequent data from the same study provided to RAND by Krista Gebert, we assumed that the cost of a large fire in our simulation would be drawn from a proportional distribution subject to two rules: The mean of the distribution is $3.3 million, and the fire with the nth percentile among all SCI scores in the simulation will have a cost that is proportional to the fire at the nth percentile of historical fire costs.

In Figure 6.1, we rank the estimated costs of large fires using the SCI approach. While the average large fire costs about $3.3 million (where we have placed the vertical axis in the figure), our approach estimates that more than half of large fires cost less than $1 million. More than 70 percent cost less than the $3.3 million average. The social costs of large fires have a long right tail, meaning that a few very costly

Figure 6.1
Cumulative Distribution of Prospective Large Fire Costs

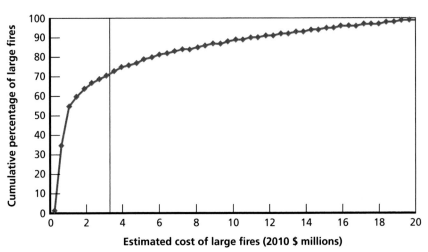

fires disproportionately account for the annual social costs attributable to fires. The $3.3 million average social cost estimate masks that heterogeneity.

There are important assumptions undergirding this procedure. For example, we assume that nascent large fires that are successfully prevented by aerial attack save social costs that are equal to those of historical fires that were not successfully prevented (i.e., that became large). However, if "aviation-vulnerable" fires are less virulent than the average observed large fire, our approach risks overvaluing air attack.

Key Unknowns Affecting Model Results

Dispatch Prescience

As in the National Model, dispatch prescience proves to be an important source of uncertainty influencing Local Resources Model outcomes. In the real world, incident commanders and fire center dispatchers make dispatch requests and decisions based on the capacity of available assets to produce fire-control lines and their perceptions of the need for and value of air support.

In the Local Resources Model's optimization, the consequences of sending or not sending an air asset are known with certainty, and dispatch can be optimized to send aircraft only to those fires where they provide maximum net benefits. Clearly, this level of dispatch prescience greatly exceeds what is feasible in real firefighting operations. Thus, we needed to degrade the Local Resources Model's prescience so that aircraft are also dispatched to a realistic number of fires where aircraft are "wasted," either because local resources alone could have contained the fire or because, even with the aid of aircraft, the fire cannot be contained.

Basing Efficiency

Because the Local Resources Model tracks aircraft locations, both when they are dispatched to fires and when they return to bases, it introduces a second level of prescience or foresight having to do with the efficiency with which aircraft can be based and pre-positioned for subsequent fires. Because the model knows with certainty where tomorrow's fires will be and which fires could benefit most from each type of aircraft, the model is capable of identifying optimal fleet mixes that imply unrealistic levels of what might be termed "basing efficiency" on the part of the Forest Service. In reality, dispatchers may not know where aircraft will be most needed tomorrow with sufficient confidence and precision to justify daily pre-positioning of aircraft. Moreover, although we did not want to constrain the model to assume that airtankers would operate only from the existing air bases currently configured to support airtankers, it is equally unreasonable to assume that airtankers will have perfect flexibility to operate out of every airport in the United States. A further limitation on perfectly efficient basing may be that aircrews might not be available to move aircraft to the optimal base after completing a full day of firefighting.

For these reasons, we allow for the possibility that aircraft basing is not perfectly efficient. Instead, we can allow for some inefficiency in basing and pre-positioning that we present in the model as inflexibility in relocating aircraft. Assuming high efficiency, we allow the model to relocate aircraft to new bases at the end of each day. With less efficiency, the model is prevented from relocating aircraft more than once every two or more days.

Water Access

The Local Resources Model makes the same water access assumptions as the National Model. For example, both models assume that a helicopter needs a body of water with a 12-foot minimum diameter, while a scooper needs a body of water with a length of at least 4,400 feet.

Retardant-Water Efficacy

As noted earlier, FPA and, hence, the Local Resources Model assume that water is as effective as retardant at building a fire chain on a per-gallon basis. The National Model makes what we believe to be the more defensible assumption that retardant is twice as effective as water on a per-gallon basis.

We did not have the ability to change FPA's water-efficacy assumption. In the National Model, as shown in Table 5.3 in Chapter Five, we found that the crossover point between favoring a scooper-centric portfolio and favoring an airtanker-centric portfolio does not occur until water is highly ineffective relative to retardant.

Results from the Local Resources Model

We performed sensitivity analyses for the factors for which limited data were available:

- basing efficiency
- dispatch prescience
- social costs of large fires.

Our approach to evaluating the model's results and placing them in their proper context, given the existing uncertainties that constrain the current analysis, was therefore to explore their sensitivity to variations in each of these key factors.

Figure 6.2 shows where aircraft of all three types are estimated to be operating in July and August, assuming high basing efficiency, impaired dispatch prescience, and an average large fire cost of $3.3 million. (July and August are the critical, highest-demand months for initial attack, though our simulation, in fact, covers June through September.) Under these assumptions, the fleet is relatively small, with just one 3,000-gallon airtanker, two 2,700-gallon helicopters, and 15 1,600-gallon scoopers. Not surprisingly, we see the highest level of aircraft use in the West, with particular concentrations in Idaho, western Montana, and Utah.

Table 6.1 highlights the sensitivity of the optimal fleet size to variations in basing efficiency. As basing efficiency diminishes (i.e., as aircraft are assigned to a given base for more days), the optimal fleet size rises; thus, when we assume that aircraft are assigned to a specific base for 20-day stretches, the fleet includes seven large helicopters, four airtankers, and 25 scoopers.

In each case, the optimal fleet is dominated by scoopers, with airtankers and helicopters used in ancillary roles in initial attack. Air-

Figure 6.2
Estimated Locations of Aircraft Usage in July and August

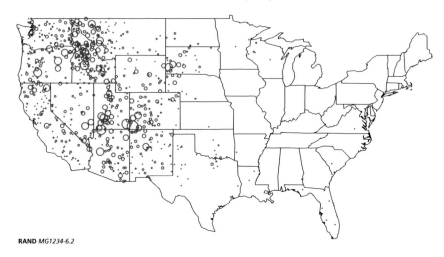

Airports serving as airbases
(circle size indicates frequency of use)

Table 6.1
Result Sensitivity to Different Minimum Basing Durations in the Local Resources Model

Minimum Number of Days Assigned to a Base	2,700-Gallon Helicopters	3,000-Gallon Airtankers	1,600-Gallon Scoopers
1	2	1	15
2	4	2	18
6	6	2	19
14	6	2	25
20	7	4	25

NOTE: Shading represents the model's base-case estimate.

tankers are sent against fires that are distant from water, while helicopters are used against fires where helicopter-accessible (but not scooper-accessible) water is available.

Table 6.2 contrasts the chosen number of aircraft associated with average large fire costs of $2.1 million, $3.3 million, and $4.5 million—our range of large fire cost estimates, discussed in Chapter Three—along with very low and high values of $300,000 and $10 million. Higher large fire costs suggests larger fleet sizes, though the magnitude of this effect is quite attenuated for large fire costs above our midrange value of $3.3 million. Indeed, more than doubling the average cost of a large fire from $4.5 million to $10 million increases the total fleet size only slightly. These results assume high basing efficiency (daily pre-positioning) and impaired prescience.

To examine the effect of different levels of dispatch prescience on optimal fleet mixes, we distinguished three cases:

- With "perfect prescience," dispatchers succeed in sending large aircraft only to Category B fires, those fires that would become large without the aircraft but are contained when the aircraft is dispatched against them.
- With "impaired prescience," dispatchers cannot distinguish between Category B and Category C fires, so they send aircraft to them on a first-come, first-served basis, as well as against a

Table 6.2
Result Sensitivity to Different Large Fire Average Costs in the Local Resources Model

Average Cost of a Large Fire	2,700-Gallon Helicopters	3,000-Gallon Airtankers	1,600-Gallon Scoopers
$0.3 million	2	1	6
$2.1 million	2	2	14
$3.3 million	2	1	15
$4.5 million	4	1	14
$10 million	5	2	16

NOTE: Shading represents the model's base-case estimate.

subset of Category A fires that have SCI scores comparable to that of a Category B or Category C fire (which we again refer to as "close-call" Category A fires). Impaired prescience is our base-case assumption.

- With "poor prescience," we assume that dispatchers are obliged to send at least one large aircraft to every Category B, Category C, and close-call Category A fire.

Table 6.3 presents the different fleet mixes implied by these prescience assumptions, assuming high basing efficiency (daily pre-positioning) and a $3.3 million average large fire cost.

Not surprisingly, poor prescience—fighting every close-call Category A fire, every Category B fire, and every Category C fire—suggests the need for a much larger fleet than do the more prescient approaches. Interestingly, our impaired prescience assumption does not lead to many fewer fires being fought than the poor prescience assumption. Indeed, more than 90 percent of all fires fought in the poor prescience scenario are also fought in the impaired prescience scenario, whereas perfect prescience results in an air attack on fewer than 70 percent of the same fires. An important implication of this finding is that nearly all fires can be fought with a moderately sized fleet, but a small number of fires (and days) require many more aircraft if the Forest Service determines that they all must be fought. It is vastly more expensive to have enough aircraft to fight all the fires on the worst day of a season than to have enough aircraft to fight all the fires on most days.

Table 6.3
Result Sensitivity to Different Dispatch Prescience Assumptions in the Local Resources Model

Prescience Assumption	2,700-Gallon Helicopters	3,000-Gallon Airtankers	1,600-Gallon Scoopers
Perfect prescience	4	2	9
Impaired prescience	2	1	15
Poor prescience	7	8	48

NOTE: Shading represents the model's base-case estimate.

The Need for Sensitivity Testing in the Local Resources Model

The dominance of scoopers in the Local Resources Model results may reflect limitations in both the FPA model's assumptions and in our selection of appropriate bodies of water. In particular, FPA assumes that one gallon of water has fire-control capabilities equal to one gallon of retardant. In the National Model, the preference for a scooper-centric fleet is preserved even when water is only a fraction as effective as retardant on a per-gallon basis. We were not able to estimate the crossover value in the Local Resources Model.

Similarly, FPA and the Local Resources Model draw no distinction between direct and indirect attack tactics. In fact, while retardant can be used in indirect attack, water typically cannot because it evaporates too quickly. Thus, both of these assumptions may exaggerate the effectiveness of the water carried by scoopers relative to the retardant carried by large airtankers. Additionally, if many of the bodies of water we identified for use by scoopers are actually inappropriate due to problems with their depth, obstructions to safe flight, or water rights, the role of scoopers would again be exaggerated in our models.

Supplementing the Initial Attack Fleet to Support Large Fire Operations

Both the National Model and the Local Resources Model analyze the type of aircraft to be used in an initial attack. In our initial attack analysis, we compared the costs of large aircraft to the costs of a large fire.

Another significant area in which large aerial firefighting assets are frequently used is fighting already-large fires. Aerial firefighting assets may be used to protect individual homes or to steer the flame front around particularly sensitive areas. Unfortunately, this is a use that has undergone comparatively little empirical study to establish the value that aircraft produce in such large fire operations. Instead, there are many anecdotes, but no statistical data have been collected to our knowledge to establish the frequency with which air attack produces benefits or the magnitude of those benefits.

Without data on the effectiveness of aircraft against large fires, we could not model the optimal mix of aircraft for large fire operations. Instead, we answer a more general question: How much value would aircraft need to produce to justify acquiring more than would already be available in a cost-effective initial attack fleet? This kind of "break-even analysis" does not assess the value of aircraft against large fires; that valuation estimate is an input to this analysis.

Our analysis assumes that the Forest Service already has the optimal number of assets needed for initial attack and that any slack capacity in these assets would be used on already-large fires prior to acquiring any additional aircraft. By examining aircraft utilization rates in the Local Resources Model, along with the modeled number and timing of large fires, we can establish the number of days of aircraft service that each large fire can receive using just the initial attack fleet. Increasing the minimum number of aircraft days available for each large fire requires the acquisition of additional aircraft beyond the initial attack fleet.

Figure 6.3 relates the number of additional scoopers that should be acquired to

- the average benefit or savings associated with each day of scooper air attack on large fires (e.g., the values at risk that might be saved each day by the use of an asset on large fires)
- the number of days of air attack that could produce such benefits.

In Figures 6.3–6.5, "1 day" refers to one day of effective use on a large fire, "2 days" refers to two days of such use, and "7 days" refers to one week of such use.

For example, if scoopers produce an average of $1 million in benefit each day they fight a large fire, and if the average large fire has at least one day when scoopers can produce that benefit, the blue line in Figure 6.3 indicates that the Forest Service would be justified in acquiring one additional scooper to cover large fire operations. However, if large fires need an average of just one day of scooper sup-

Figure 6.3
**Desirability of Acquiring Extra 1,600-Gallon Scoopers for Use Against
Already-Large Fires**

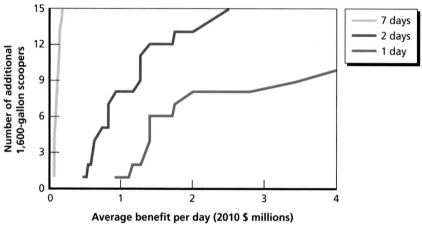

RAND *MG1234-6.3*

port and the average benefit of a scooper is anything less than about $1 million per day, using slack capacity in the existing initial attack fleet would be the most cost-effective approach to meeting large fire aerial attack requirements.

In contrast, if the average large fire can benefit from seven days of scooper support (the green line in Figure 6.3), and if the average benefit on each day of operations is even just $100,000, the Forest Service would be justified in acquiring seven additional scoopers to complement its initial attack fleet.

Figures 6.4 and 6.5 are analogous figures for 3,000-gallon airtankers and 2,700-gallon helicopters.

The results in Figures 6.3–6.5 build on the fairly small baseline Local Resources Model initial attack fleet, based on the assumptions of high-efficiency basing, impaired prescience, and a $3.3 million average large fire cost. If initial attack fleets are larger, fewer additional aircraft would be justified to cover large fire operations, as the initial attack fleet would, on average, have a lower utilization rate.

The analysis that generated Figures 6.3–6.5 assumed that only one type of large aircraft would be acquired. If, instead, the Forest Ser-

**Figure 6.4
Desirability of Acquiring Extra 3,000-Gallon Airtankers for Use Against
Already-Large Fires**

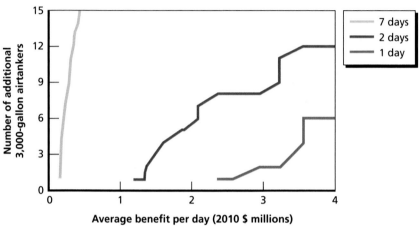

RAND *MG1234-6.4*

**Figure 6.5
Desirability of Acquiring Extra 2,700-Gallon Helicopters for Use Against
Already-Large Fires**

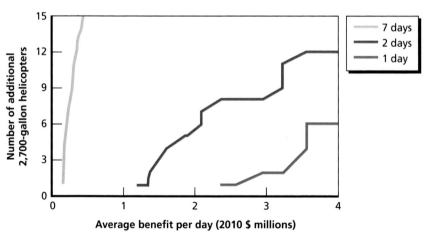

RAND *MG1234-6.5*

vice decided to acquire a mix of aircraft to support large fires, a new breakeven analysis that blends information about each aircraft type would be needed.

Using These Results

The Local Resources Model provides a baseline recommendation for two 2,700-gallon helicopters, one 3,000-gallon fixed-wing airtanker, and 15 1,600-gallon scoopers to be used in initial attack operations. This fleet assumes a daily adjustment to aircraft basing, only slight impairments to the prescience of dispatchers, and that water is equally as effective as retardant on a per-gallon basis. Furthermore, this baseline recommendation assumes that no additional large aircraft are required to fight already-large fires. Each of these assumptions has a substantive impact on the optimal fleet of aircraft, but the firefighting literature does not have conclusive estimates for many of the key parameters. In this section, we discuss how to use the baseline recommendation, uncertainty in the assumptions, and the requirements for large fires to determine a final target for the Forest Service's fleet. Complementing this discussion, Appendix B discusses trends in fire aviation demand through 2030.

The baseline recommendation assumes that aircraft can be repositioned on a daily basis in response to expected fire conditions. This could be too optimistic because of a lack of information, poor coordination of resources, or operational requirements resulting from, for instance, crew availabilities and shift schedules. The less efficiently these aircraft are based, the more aircraft are needed to cover the area at risk. Table 6.1 indicates that if aircraft are rebased weekly, four additional helicopters and four additional scoopers would be needed to cover the area at risk. If aircraft are rebased only every other week, four additional helicopters and ten additional scoopers would be required. This highlights the importance of aircraft deployment: The Forest Service could accrue substantial savings if the deployment practices were optimized.

In addition to the degree of imperfection in the basing, the degree to which fires are optimally targeted for initial attack by aircraft has a substantial impact on the need for aircraft. Dispatchers have imperfect information about whether the aircraft attack will make a difference in the small-to-large fire outcome. That is to say, we do not know how often aircraft are sent to fires that would have been contained with local resources alone or that would have become large regardless of the addition of air resources. We tested levels of efficiency in the dispatch prescience. As shown in Table 6.3, perfect fire selection results in fewer scoopers in the optimal fleet, while poor dispatch prescience results in a substantially higher requirement for each type of aircraft. Dispatch efficiency is clearly a very important parameter. Improving the intelligence available to dispatchers can dramatically reduce the number of aircraft required and increase the efficiency of their use.

In addition to using aircraft for initial attack, there may be a desire to use aircraft against already-large fires. There is no consensus on how much, if any, benefit large aircraft provide against already-large fires. Despite the lack of information, the use of aircraft against large fires cannot be ignored because they are often used in this role. With this in mind, we determined how many large fires could not be attacked using the baseline fleet and then calculated how much value an aircraft would need to provide to justify its use against large fires only. Figures 6.3–6.5 present curves indicating the how much value each aircraft would need to provide to justify acquiring additional aircraft. So, for example, if a scooper could save an average of $1 million in property damage for two days per large fire, the Forest Service should have seven additional scoopers for use against large fires. We found that airtankers and helicopters would require much higher assumptions about the savings per day.

Putting these results together, we can produce a template for an optimal fleet. The bulk of the analysis indicates that scoopers should be the primary aircraft in the fleet. There are some parts of the country that are poorly served by scoopers because of water availability problems, so some airtankers will be needed. Likewise, helicopters can fill a specific niche where available water bodies make scoopers less useful. Table 6.4 shows the range of reasonable fleet mixes under this uncer-

Table 6.4
Illustration of Suggested Fleet Sizes in the Local Resources Model

Type of Aircraft	Baseline Fleet	Basing Inefficiency	Dispatch Inefficiency	Large Fire	Total Aircraft
Airtankers	1	0 to +2	0 to +3	0	1–4
Helicopters	2	0	0 to +6	0	2–8
Scoopers	15	0 to +10	0 to +30	+7	15–45

NOTE: The columns in the table are not additive. For instance, if one adds ten additional scoopers to address basing inefficiency, one would not need to add 30 additional scoopers to address dispatch inefficiency.

tainty. Further study and expert judgment should be used to narrow these ranges and select a final fleet.

The primary caveat here is that a scooper-centric fleet is preferred so long as the model's assumptions about the efficacy of water relative to retardant are reasonable. In the National Model (see Chapter Five), a scooper-centric fleet is generally preferred over an airtanker-centric one, but that gap narrows if water produces less than one-third the chains per gallon as retardant. From our discussions with Forest Service personnel and others, we believe that water should be considered half as effective as retardant, so a scooper-centric fleet is preferable, but the results of this analysis are dependent on this assumption.

CHAPTER SEVEN

Concluding Remarks

The two models we developed to assess the mix of large aircraft that would optimize society's returns on investment in Forest Service initial attack capabilities provided a frustratingly broad range of answers—from 18 large aircraft in the Local Resources Model to 56 large aircraft in the National Model GACC-restricted variant (see Table 7.1).[1]

As Table 7.1 shows, the National Model suggests an increase in the number of scoopers (from 40 to 43) when water goes from efficacy parity with retardant to being half as effective. Further, the National Model again recommends 43 scoopers when water is one-quarter as effective as retardant. The reason for this counterintuitive result is that reducing the assumption about water's efficacy to half that of retardant, more scoopers are needed (along with more airtankers) to offset the diminution in capability associated with any single scooper. As we showed in Table 5.3 in Chapter Five, it is only when water is highly ineffective that there is a considerable diminution in demand for scoopers.

Unfortunately, we were not able to run a parallel water-efficacy diminution exploration using the Local Resources Model because water-retardant efficacy parity was intrinsic to the FPA model. Therefore, we have placed an "X" in the two reduced water-efficacy cells under the Local Resources Model in Table 7.1.

---

[1] Table 7.1 displays results for an initial attack fleet only. A more complete account of Forest Service aircraft requirements would include aircraft for fighting already-large fires. Chapter Six discusses how a total fleet size and mix can be constructed to include aircraft for both initial attack and use against already-large fires. However, it would be conditional on having an estimate of the daily value of aircraft against already-large fires.

Table 7.1
Estimates of Optimal Initial Attack Fleets in the National and Local Resources Models

Case	RAND National Model	RAND Local Resources Model
Water-retardant efficacy parity	2 airtankers, 40 scoopers	1 airtanker, 15 scoopers, 2 helicopters
Water half as effective as retardant	5 airtankers, 43 scoopers	X
Water one-quarter as effective as retardant	9 airtankers, 43 scoopers	X
$2.1 million per average large fire	4 airtankers, 36 scoopers	2 airtankers, 14 scoopers, 2 helicopters
$4.5 million per average large fire	6 airtankers, 55 scoopers	1 airtanker, 14 scoopers, 4 helicopters
Geographical constraint	8 airtankers, 48 scoopers	4 airtankers, 25 scoopers, 7 helicopters

NOTE: The National Model geographical constraint case restricts aircraft to operating in a single GACC in a given month. The Local Resources Model geographical constraint case has each aircraft assigned to a given base for at least 20 days. The Local Resources Model does not allow varying water efficacy. The shaded cells represent each model's base-case estimate.

Much of the variability in our models' results derives from fundamental uncertainties in the science and economics of wildland firefighting, such as the relative effectiveness of water and retardant, the value of fighting large fires, the efficiency of aircraft dispatch and prepositioning, the true social costs of large fires, the likely future severity of fire seasons, and even the true efficacy of air attack against wildfires.

Despite these uncertainties, our modeling efforts and the results across a wide range of different assumptions provide important insights that can assist the Forest Service in determining its recapitalization needs. In this concluding chapter, we highlight what we believe are some of the key insights suggested by our analyses.

Across our analyses, scoopers were found to be the central component of the optimal solution. Two factors drive this finding. First, scoopers are considerably less expensive to own and operate than larger helicopters and fixed-wing airtankers. Whereas we estimate annual scooper costs to be around $2.8 million, 3,000-gallon commercial or military fixed-wing airtankers and 2,700-gallon helicopters cost more than $7 million per year. In other words, for each pair of 2,700-plus-gallon aircraft, the Forest Service could operate more than five scoopers.

The second factor driving the optimal fleet mix toward scoopers is the volume of water that they can drop compared to the retardant drops made by airtankers. When fires are proximate to water sources—and we found that roughly 60 percent of all fires are within ten miles of a potentially scooper-accessible body of water—scoopers can drop 1,600 gallons of water every 13 minutes or less, compared with average airtanker mission times of 45 minutes. Indeed, the fires with the highest values at risk tend to be proximate to water sources precisely because most human settlement is proximate to water.

Importantly, however, water and retardant are not equally effective, nor can they be used in all the same circumstances. For instance, an important airtanker tactic is indirect attack, or dropping retardant at strategic locations in the path of the burning edge of a fire. Water cannot be used for indirect attack because it evaporates too quickly. Because a direct attack may be substantially less effective for some fires than an indirect attack, we believe that retardant must have greater effectiveness in supporting fire-control line building. But our robustness explorations suggest that a scooper-centric portfolio is preferred for plausible levels of water efficacy.

We learned that several U.S. states, Canadian provinces, and European countries have become heavily reliant on scoopers for wildland firefighting. We contacted many of these users and heard universally positive reviews of scooper efficacy, suggesting that their inability to conduct indirect attacks does not result in a dramatic loss in firefighting efficacy. Retardant-bearing airtankers are also valuable, but primarily in the niche role of fighting the minority of fires that are not water-proximate.

In developing both the National Model and the Local Resources Model, we confronted the issue of dispatch prescience and its importance in determining optimal initial attack aircraft portfolios. Dispatch prescience is both strategic (how skillfully and flexibly aircraft are prepositioned) and tactical (which small fires on a given day are selected for attack by aviation). Our analyses demonstrate that greater dispatch prescience can sharply reduce the size of required initial attack aircraft portfolios. Indeed, model results are sufficiently sensitive to dispatch prescience that we were obliged to construct crude estimates of the level of dispatch prescience currently exhibited by dispatchers. From our conversations with Forest Service fire and aviation managers, we hypothesized that a plausible level of dispatch prescience might be one that makes no distinction between Category B fires (which can be successfully attacked by aircraft) and Category C fires (which cannot). Moreover, we added close-call Category A fires, assuming that dispatchers and incident commanders lack sufficiently good intelligence on fires to distinguish them from Category B fires.

Because dispatch prescience and pre-positioning proved to be so important to the optimal fleet mix, we recommend that the Forest Service conduct focused studies examining the success of current dispatch and pre-positioning efforts to match air attack assets to Category B fires. The Forest Service may be able to substantially reduce the costs of fighting fires with improved fire intelligence. By improving the information and, perhaps, the science base used by dispatchers to decide which fires require aircraft, the Forest Service could substantially reduce the number and cost of its large aircraft fleet.

A related useful finding from our analysis of dispatch prescience is that a fleet that is optimally sized to attack Category B, Category C, and close-call Category A fires would be able to fight more than 90 percent of all such fires. Nearly all fires can be fought with a moderately sized fleet, but a small number of days require many more aircraft if the Forest Service determines that all such fires all must be fought. Thus, it is vastly more expensive to have enough aircraft to fight all fires on the worst day of a season than to have enough aircraft to fight all the fires on most days.

An additional area for further inquiry is the value of aircraft against already-large fires. Table 7.1's fleets are for initial attack only. While some of those aircraft might be used on an as-available basis against already-large fires, the perhaps-unrealistic assumption is that they would be diverted as needed to their assumed primary mission, initial attack.

Given the frequency with which large airtankers are used against already-large fires, we were surprised by the dearth of statistical evidence documenting their value in this role. Better information about the costs and benefits of air assault in large fire operations will help clarify the optimal mix of aircraft required for wildland firefighting operations. Until that information becomes available, the Forest Service may have no better means of determining the number of additional aircraft it requires beyond those required for initial attack than the breakeven analyses presented in Chapter Six. These analyses require estimates of daily benefits provided by aircraft that are dispatched against large fires.

This study considered factors influencing the number and mix of large aircraft that would optimize taxpayer returns on investment. We did not explore the range of acquisition strategies that might be considered to achieve the needed fleet mix. A detailed acquisition strategy will be needed that considers the rate at which new aircraft are acquired, whether they will be purchased or leased, and whether some peak demands could be met using military-operated aircraft.

Along with an evaluation of acquisition strategies, there are several other possible extensions of this work, including

- allowing water-efficacy parameter flexibility in FPA and, hence, in the Local Resources Model
- bringing local resource data from FPA into the National Model
- analyzing how Forest Service aircraft have been used, e.g., their patterns of relocation (where and how frequently), the amount of time they spend fighting small versus already-large fires
- assessing, perhaps experimentally, how often aircraft truly change outcomes between small and large fires
- calibrating the frequency and efficacy of direct versus indirect attack in today's airtanker fleet.

Any of these extensions would abet further efforts to understand the Forest Service's requirements for large aircraft.

Equations Used to Construct High and Low Fire Cost Estimates

This appendix presents the formal analyses underlying the large fire cost estimates in Table 3.3 in Chapter Three.

Federal Fire Suppression Costs

D = \$150,000, the suppression costs of a DOI large (100-plus-acre) fire from Table 3.4 (\$531,747,000 / 3,550).

F_{100} = \$3.6 million, the average suppression cost of a Forest Service (100-plus-acre) fire from Table 3.4 (\$5,628,201,000 / 1,542).

F_{300} = \$5.6 million, the average suppression cost of a Forest Service (300-plus-acre) fire from Table 3.4 ([\$605,469,000 + \$1,528,205,000 + \$3,394,944,000] / [373 + 362 + 244]).

G_{300} = \$3.6 million, a lower estimate of the average suppression cost of a Forest Service (300-plus-acre) fire from Gebert's data (Gebert, 2011).

$G_{100} = G_{300} \times F_{100} / F_{300}$ = \$2.3 million, our scaling of the Gebert estimate to arrive at an approximate cost of a 100-plus-acre fire (see the discussion of Table 3.4 in Chapter Three).

P_{H} = 0.9, a high estimate of the proportion of fires on Forest Service land that are fought by Forest Service large airtankers (see the discussion of Table 3.4 in Chapter Three).

P_L = 0.8, a low estimate of the proportion of fires on Forest Service land that are fought by Forest Service large airtankers (see the discussion of Table 3.4 in Chapter Three).

$H = P_H \times F_{100} + (1 - P_H)D$ = \$3.3 million, the high estimate of federal suppression costs in Table 3.3 (the federal fire suppression high).

$L = P_L \times G_{100} + (1 - P_L)D$ = \$1.9 million, the low estimate of federal suppression costs in Table 3.3 (the federal fire suppression low).

State and Local Fire Suppression Costs

S = 0.07, Gebert and Schuster's (2008) estimate of the proportion of total suppression expenditures paid by state and local agencies (see the section "State and Local Fire Suppression Costs" in Chapter Three).

$S / (1 - S)$, the proportion of Forest Service suppression costs equivalent to state and local costs (0.07 / 0.93).

$L_H = H(S / [1 - S])$ = \$248,000, the high estimate for the state and local costs in Table 3.3 (see the section "State and Local Fire Suppression Costs" in Chapter Three).

$L_L = L(S / [1 - S])$ = \$142,000, the low estimate for the state and local costs in Table 3.3 (see the section "State and Local Fire Suppression Costs" in Chapter Three).

Large Aircraft Suppression Costs

A_P = 0.16, the estimate of the proportion of large fire costs attributable to the use of all aviation (see the section "Large Aircraft Costs" in Chapter Three).

A_L = 0.83, the proportion of Forest Service aviation expenses attributable to large aircraft, according to Brosnan (2008) (see the section "Large Aircraft Costs" in Chapter Three).

$A_p \times A_L \times L$ = \$250,000, the estimated costs of large aviation deducted from the low estimate in Table 3.3 (a low estimate of \$1.879 million in federal fire suppression costs per fire multiplied by 0.16 and by 0.83).

Small Fire Suppression Costs

From Table 3.4, average small fire (less than 100 acres) costs are \$7,000 ([\$5,912,567,000 − \$5,628,201,000] / [41,059 − 1,542]).

Rehabilitation of Burned Lands

R = \$77 million, burned-land rehabilitation expenses reported by the Forest Service from 2005 to 2009 (U.S. Department of Agriculture, Forest Service, 2005, 2007, 2009, 2010; see the section "Rehabilitation of Burned Lands" in Chapter Three).

L_N = 1,542, the number of large fires over the same period, from Table 3.4 (the Forest Service large fires cell in the lower left corner of Table 3.4).

R / L_N = \$50,000, the estimate of the average rehabilitation costs for large fires appearing in Table 3.3 (\$77 million divided by 1,542).

Insured Losses

U = 10,000, Gude et al.'s (2009) estimate of the number of housing units lost to wildfires between 2002 and 2006 (see the section "Insured Losses" in Chapter Three).

V_U = \$193,374, the estimated replacement cost of houses in the Western United States during this period (from Davis and Heathcote, 2007; see the section "Insured Losses" in Chapter Three).

$V_L = U \times V_U / 5$ = \$387 million, the low estimate for annual replacement costs for housing lost to wildland fire (10,000 × \$193,374 / 5).

I_V = $905 million, annual insured losses attributed to wildfires between 2000 and 2007, according to the International Code Council (2008) (see the section "Insured Losses" in Chapter Three).

N = 1,177, estimate of the total number of large fires in the United States, including federal, state, and other fires (estimate from NIFC wildfire statistical reports [National Interagency Coordination Center, 2005–2009]; see the section "Insured Losses" in Chapter Three).

V_L / N = $329,000, the low estimate of structural losses attributable to the average large wildfire in Table 3.3 ($387 million / 1,177).

I_V / N = $769,000, the high estimate of structural losses attributable to the average large wildfire in Table 3.3 ($905 million / 1,177).

Loss of Life

F = 173, the number of wildland firefighter fatalities between 1999 and 2006 (National Wildfire Coordinating Group, 2007; see the section "Loss of Life" in Chapter Three).

P_L = 0.88, the proportion of the 47.2 million acres burned in U.S. wildfires from 2004 to 2009 that can be attributed to fires over 100 acres (calculated from FIRESTAT database).

$F_A = P_L \times F / (N \times 8 \text{ years})$ = 0.0162, annual wildland firefighting fatalities attributable to large fires (0.88 × 173 / [1,177 × 8])

S_L = $3.75 million, low estimate of the value of a statistical life saved (Viscusi and Aldy, 2003).

S_H = $7.87 million, high estimate of the value of a statistical life saved (Viscusi and Aldy, 2003).

$S_L \times F_A$ = $61,000, low estimate of average fatality losses attributable to large fires in Table 3.3 ($3.75 million × 0.0162).

$S_H \times F_A$ = $127,000, high estimate of average fatality losses attributable to large fires in Table 3.3 ($7.87 million × 0.0162).

Estimating Future Suppression Cost Savings of a Large Fire

One of the most uncertain calculations we undertook was to estimate the future cost savings associated with a large fire. Presumably, after a large fire, an area will be less vulnerable to a large fire in the future. However, as noted in Chapter Three, the magnitude of the future risk diminution is highly uncertain. Nevertheless, we experimented with future suppression cost calculations, if only to illustrate how different beliefs about risk reduction can lead to considerably different cost estimates.

To illustrate how our future suppression calculation works, we present our analytically easier case: For three years after a large fire, assume that any subsequent fire will have only 70 percent of the cost as in the baseline case. Let p be the (assumed-to-be-unchanged) annual probability of a fire in the area, and let D denote the damages that would typically occur if a large fire hits the area.

Assume that the area has not had a fire in the past three years. Then, with probability p, a fire will occur this year and will cost D. If that happens, the affected area enters three years of 70-percent (rather than 100-percent) fire damages. With probability $1 - p$, no fire occurs next year, and the process remains in its steady state with a long-run cost that we label V. Hence,

$$V = p \times D + \frac{p \times R_3}{1+i} + \frac{(1-p) \times V}{1+i}, \text{ or}$$

$$V = \frac{p \times D + \dfrac{p \times R_3}{1+i}}{(1 - \dfrac{(1-p)}{1+i})},$$

where i is the long-term real interest rate and R_3 denotes the expected cost of being in the first of three years of reduced fire cost. Likewise, R_2 denotes the expected cost of being in the second of three years of reduced fire cost, while R_1 denotes the expected cost of being in the last of three years of reduced fire cost. (R's subscript denotes the years remaining in the reduced-cost state.)

Starting in state R_3 (i.e., a fire occurred last year), the parallel equation would be

$$R_3 = (0.7) \times p \times D + \frac{p \times R_3}{1+i} + \frac{(1-p) \times R_2}{1+i}, \text{ or}$$

$$R_3 = \frac{0.7 \times p \times D + \frac{(1-p) \times R_2}{1+i}}{1 - \frac{p}{1+i}}.$$

By assumption, damages from a fire are 70 percent as large as the baseline if there was a fire in the past three years.

Likewise, we have

$$R_2 = 0.7 \times p \times D + \frac{p \times R_3}{1+i} + \frac{(1-p) \times R_1}{1+i} \text{ and}$$

$$R_1 = 0.7 \times p \times D + \frac{p \times R_3}{1+i} + \frac{(1-p) \times V}{1+i}.$$

After three years in the "protected" status with 70-percent fire damages, if there is no fire, the process returns to its long-run steady state with cost V.

To estimate p, the baseline probability of a fire, we note that approximately 0.5 percent of the Forest Service's 188 million acres burned annually over the period 2004–2009, so we let $p = 0.005$. OMB (2009) prescribed a 2.7-percent long-term real interest rate for 2010, so we have $i = 0.027$. $D = \$4.470$ million is the sum of all of our high cost categories, except future suppression. Then, we find that $V = 848,837$, $R_3 = 829,160$, $R_2 = 835,513$, and $R_1 = 842,069$.

Suppose a successful initial attack occurs. Successful initial attack implies that instead of being in state R_3 next year, one is in state V. The present value of this change in status is

$$\frac{V - R_3}{1 + i},$$

which equals $19,159. We therefore estimate a successful initial attack-induced increase in future suppression costs equal to about $19,000 in our high case in Table 3.2.

Our low case is analytically more complicated, though conceptually similar. Now there are 13 equations (V and $R_{12}...R_1$). The fact there are 12 years of reduced damages and those reduced damages are now assumed to be 30 percent, rather than 70 percent, of the baseline value more than offsets the fact that D = $2.205 million in the low case. The result is a successful initial attack-induced increase in future suppression costs equal to about $76,000 in our low case in Table 3.2.

We believe that further research is needed on how current fires change future fire probability and severity.

Although we have presented a technique for considering post-fire severity changes, we do not have a high degree of confidence in our future suppression credit estimates.

Trends in Fire Aviation Demand Through 2030

The dramatic increases in recent wildfire suppression costs, discussed in Chapter Three have been attributed to three primary factors: the accumulation of forest fuels resulting from years of successful suppression, aggressive suppression efforts to fight fires that threaten increasing numbers of homes in the WUI, and climate variation, which has produced long, hot, and dry fire seasons.

Over the next 20–30 years, growth in the WUI is almost certain to continue as a trend, producing increasing demands on firefighting resources. Fuel accumulation is likewise likely to continue. Climate change could worsen fire season severity, though this factor is less straightforward to predict.

WUI Trends

If growth in the number of housing units in the WUI proceeds at the pace it has in the past two decades, the 2000–2030 period could see 111-percent growth in WUI housing in the West and 93-percent growth in the Southeast; these two regions accounted for 91 percent of acres burned in the 48 contiguous states in 1997 (Hammer et al., 2009). This doubling of private property at risk could have a considerable effect on fire suppression costs and the demand for Forest Service aircraft.

Although protecting private property is not necessarily the Forest Service's responsibility, public pressure to assist in such efforts and to manage wildfires that threaten the WUI is intense. In a recent audit of the cost of large fires, the U.S. Department of Agriculture's Office of the Inspector General (2006) found that, in 87 percent of fires, protection of private property was listed as the key motivator for firefighting efforts. The same report describes a 1994 National Fire Protection Administration study that found that one-third of the federal fire suppression budget goes to protecting the WUI, and Forest Service staff have estimated that between 50 percent and 95 percent of their suppression expenses go to protecting private land and homes in the WUI.

These impressions have been supported by econometric analyses of the factors affecting large fire suppression costs (Gebert et al., 2008; Liang et al., 2008). For example, Gebert et al. (2008) found that, other factors being equal, for each percentage point increase in housing value per acre in the 20-mile radius of a fire's ignition point, there is a 0.1-percent increase in expected suppression costs. Using this relationship, we considered the expected increase in suppression costs from 2008 to 2030 if the number of housing units in the WUI were to increase by 75 percent (roughly corresponding to the projected doubling of units from 2000 to 2030; Hammer et al., 2009). In 2009, the average value of housing within 20 miles of each of the 78,512 ignitions in our historical fire data set was $5.6 billion, according to our analysis of census data. Gebert et al.'s (2008) parameter estimate suggests that growth in the WUI will lead to suppression costs that are 7.4 percent higher than current costs. Assuming that housing losses and rehabilitation costs also double with this growth, our respective 2030 high and low estimates for the value of preventing large fires rise to $5.4 million and $2.6 million in 2010 dollars.

Note, however, that the relationship between the value of averting fires and the optimal number of large aircraft is quite insensitive to large fire costs above $5 million (see Table 6.2 in Chapter Six). In other words, even if the value of averting fires increased markedly, it does not appear that there would be a commensurate increase in the optimal size of the Forest Service's initial attack aviation fleet.

Others have come to quite different conclusions. Examining projections for development in Montana's WUI, for instance, Gude et al. (2009) predicted that firefighting costs would rise 72 percent by 2025. Adding a modest climate change assumption that average fire season temperatures in Montana could increase by one degree Fahrenheit by 2025, Gude et al. (2009) calculated that wildland firefighting costs in Montana would double or even quadruple under current policies. A doubling of suppression costs, along with a doubling of housing and rehabilitation costs, would raise our respective high and low estimates of the cost of preventing large fires to $10.8 million and $5.2 million. The quadrupling of suppression costs brings our estimates to $21.6 million and $10.4 million, respectively. Again, however, the effects on the optimal size of the Forest Service initial attack fleet may be minimal through most of this range.

Fuel Accumulation

Since at least the 1995 Federal Wildland Fire Policy, U.S. fire policy has emphasized the economic and ecological benefits of wildland fire and the importance of allowing fires to burn when this can be done safely. The 1995 policy called for burnable regions to develop fire management plans to carefully reintroduce fire and reduce fire return intervals. The required plans were supposed to offer a proactive and cost-effective way to reduce the risk of catastrophic fires through the management of fuels and prescribed and unintentional fires.

By requiring management plans at the local level, the policy recognized that the costs and benefits of fire suppression are highly variable, depending on local values, property at risk, and other factors. But at the same time, it pushed complex (and contentious) analytical problems to local and regional managers, such as how to value the long-term benefits of ecological improvements, how to evaluate the trade-offs between ecological benefits and property damage, how to determine the relationship between the harm of a fire now and the risk of a potentially more serious fire later, and how to assess a fire's effect on air and water quality, along with the need to consider less tangible costs—to wildlife, culturally important sites, forest aesthetics, and so on. The challenge that this created for regional and local fire managers was further compounded by the dearth of scientific and policy guidance on how to estimate these costs and benefits (Hesseln and Rideout, 1999).

The result of this arrangement, which continues today, is that nearly all fires are targeted for suppression rather than being permitted to burn as "wildland fire use" fires. Between 1998 and 2008, the average number of "fire use" fires per year was 327, or 0.4 percent of wildfires. These fires burned 187,416 acres, or about 2.8 percent of total acres burned during those years (National Interagency Coordination Center, 2009).

In contrast, the use of prescribed fires or controlled burns has succeeded in burning approximately 2 million acres per year over the past decade, a land area equal to about one-quarter of the federal land area burned by wildfires annually. However, prescribed fires are controversial and can face stiff public opposition that can make them all but impossible to conduct near the WUI areas where they might be most beneficial. Moreover, even if they could be used at will, at 2 million acres per year, it would take 65 years to return fire to the 131 million acres of the National Forest System that has been found to be at moderate or severe risk of catastrophic wildfire due to fuels accumulation. This period too lengthy to achieve net reductions in fuel accumulations (U.S. Department of Agriculture, Office of the Inspector General, 2009).

We conclude that unless there are significant changes in fuel management policies, the trend of increasing fire severity resulting from fuel accumulation is likely to continue through the year 2030.

Climate Trends

Examining Forest Service suppression expenditures between 1970 and 2002, Calkin et al. (2005) present evidence that costs per acre burned have been relatively stable even as total wildland firefighting costs have soared. Their analysis suggests that the main cost driver has been increases in acres burned, which they attribute to long-term changes in weather patterns. In particular, since the 1990s, much of the western United States has experienced drought conditions believed to occur cyclically and expected to last 30 years (National Wildfire Coordinating Group, 2009).

In addition to weather cycles, climate change could affect wildfire frequency, severity, and the length of fire seasons, though these effects are likely to be less significant over the next 20 years than those associated with the current drought cycle. Models of the effects of climate change on the United States performed by the Intergovernmental Panel on Climate Change suggest that by 2080–2099, the western states may experience average summer temperatures 3–5 degrees (Celsius) warmer than temperatures recorded between 1980 and 1990, and the peak fire months of June, July, and August may be drier as well (Christensen et al., 2007).

If such trends occur, there could be an appreciable effect on fire seasons. Gude et al. (2009), using historical data on fire suppression and climate variation in Montana, found that for every degree (Fahrenheit) increase in average spring and summer temperatures, there is a 305-percent increase in area burned. Such increases in fire size and number could have a significant effect on the optimal number of Forest Service fire fighting aircraft.

References

Abt, Karen L., Jeffrey P. Prestemon, and Krista Gebert, "Forecasting Wildfire Suppression Expenditures for the United States Forest Service," in Thomas P. Holmes, Jeffrey P. Prestemon, and Karen L. Abt, eds., *The Economics of Forest Disturbances: Wildfires, Storms, and Invasive Species*, London: Springer, 2008, pp. 341–360.

Agee, James K., "Wildfire in the Pacific West: A Brief History and Implications for the Future," in *Proceedings of the Symposium on Fire and Watershed Management*, Berkeley, Calif.: U.S. Department of Agriculture, Forest Service, Pacific Southwest Forest and Range Experiment Station, General Technical Report PSW-109, 1989, pp. 11–16. As of July 2, 2012:
http://www.fs.fed.us/psw/publications/documents/psw_gtr109/psw_gtr109.pdf

Albini, F. A., G. N. Korovin, and E. H. Gorovaya, *Mathematical Analysis of Forest Fire Suppression*, Ogden, Utah: U.S. Department of Agriculture, Forest Service, Research Paper INT-207, 1978.

Andrews, Patricia L., *BehavePlus Fire Modeling System, Version 5.0: Variables*, Ft. Collins, Colo.: U.S. Department of Agriculture, Forest Service, Rocky Mountain Research Station, General Technical Report RMRS-GTR-213WWW Revised, September 2009. As of July 2, 2012:
http://www.fs.fed.us/rm/pubs/rmrs_gtr213.pdf

Blue Ribbon Panel on Aerial Firefighting, *Federal Aerial Firefighting: Assessing Safety and Effectiveness*, U.S. Department of Agriculture, Forest Service, and U.S. Department of the Interior, Bureau of Land Management, December 2002. As of July 2, 2012:
http://www.fs.fed.us/fire/publications/aviation/fed_aerial_ff_assessing_safety_effectivenss_brp_2002.pdf

Boychuk, Dennis, and David L. Martell, "A Markov Chain Model for Evaluating Seasonal Forest Fire Fighter Requirements," *Forest Science*, Vol. 34, No. 3, September 1988, pp. 647–661.

Bradstock, R., J. Sanders, and A. Tegart, "Short-Term Effects on the Foliage at a Eucalypt Forest After an Aerial Application of a Chemical Fire Retardant," *Australian Forestry*, Vol. 50, No. 2, 1987, pp. 71–80.

Brosnan, Larry, U.S. Department of Agriculture, Forest Service, "Cost-Benefit Analysis for Federal Firefighting Airtankers," briefing, 2008.

Brown, Arthur Allen, and Kenneth Pickett Davis, *Forest Fire: Control and Use*, New York: McGraw-Hill, 1973.

Butry, David T., D. Evan Mercer, Jeffrey P. Prestemon, John M. Pye, and Thomas P. Holmes, "What Is the Price of Catastrophic Wildfire?" *Journal of Forestry*, Vol. 99, No. 11, November 2001, pp. 9–17.

California Department of Forestry and Fire Protection, "Air Program," web page, undated. As of July 2, 2012:
http://www.fire.ca.gov/fire_protection/fire_protection_air_program.php

Calkin David E., Krista M. Gebert, J. Greg Jones, and Ronald P. Neilson, "Forest Service Large Fire Area Burned and Suppression Expenditure Trends, 1970–2002," *Journal of Forestry*, Vol. 103, No. 4, June 2005, pp. 179–183.

Calkin, David E., Kevin D. Hyde, Peter R. Robichaud, J. Greg Jones, Louise E. Ashmun, and Dan Loeffler, *Assessing Post-Fire Values-at-Risk with a New Calculation Tool*, Ft. Collins, Colo.: U.S. Department of Agriculture, Forest Service, Rocky Mountain Research Station, General Technical Report RMRS-GTR-205, December 2007. As of July 2, 2012:
http://www.fs.fed.us/rm/pubs/rmrs_gtr205.pdf

Cart, Julie, and Bettina Boxall, "Air Tanker Drops in Wildfires Are Often Just for Show," *Los Angeles Times*, July 29, 2008.

Christensen, Jens Hesselbjerg, Bruce Hewitson, Aristita Busuioc, Anthony Chen, Xuejie Gao, Isaac Held, Richard Jones, Rupa Kumar Kolli, Won-Tae Kwon, René Laprise, Victor Magaña Rueda, Linda Mearns, Claudio G. Menendez, Jouni Raisanen, Arnette Rinke, Abdoulaye Sarr, and Penny Whetton, "2007: Regional Climate Projections," in Susan Solomon, Dahe Qin, Martin Manning, Melinda Marquis, Kristen Averyt, Melinda M. B. Tignor, Henry LeRoy Miller, Jr., and Zhenlin Chen, eds., *Climate Change 2007: The Physical Science Basis: Contribution of Working Group I to the Fourth Assessment Report of the Intergovernmental Panel on Climate Change*, Cambridge, UK: Cambridge University Press, 2007.

Conklin & De Decker Associates, *Lifecycle and Budget Costs—Jets*, Orleans, Mass., 2009a.

———, *Lifecycle and Budget Costs—Turboprops*, Orleans, Mass., 2009b.

———, *Lifecycle and Budget Costs—Helicopters*, Orleans, Mass., 2010.

Countryman, Clive M., *Use of Air Tankers Pays Off: A Case Study*, U.S. Department of Agriculture, Forest Service, Research Note PSW-188, 1969.

Dalton, Patricia, U.S. Government Accountability Office, *Wildland Fire Management: Federal Agencies Have Taken Important Steps Forward, but Additional Action Is Needed to Address Remaining Challenges*, testimony before the Committee on Energy and Natural Resources, Washington, D.C., GAO-09-906T, 2009.

Davis, Morris A., and Jonathan Heathcote, "The Price and Quantity of Residential Land in the United States," *Journal of Monetary Economics*, Vol. 54, No. 8, November 2007, pp. 2595–2620.

Dunn, Alex E., with Armando Gonzalez-Caban and Karen Solari, *The Old, Grand Prix, and Padua Wildfires: How Much Did These Fires Really Cost?* Riverside, Calif.: U.S. Department of Agriculture, Forest Service, Pacific Southwest Research Station, 2003. As of July 2, 2012:
http://www.fs.fed.us/psw/publications/2003_wildfires_report.pdf

Englin, Jeffrey, Thomas P. Holmes, and Janet Lutz, "Wildfire and the Economic Value of Wilderness Recreation," in Thomas P. Holmes, Jeffrey P. Prestemon, and Karen L. Abt, eds., *The Economics of Forest Disturbances: Wildfires, Storms, and Invasive Species*, London: Springer, 2008, pp. 191–208.

Esri, Inc., "USA Detailed Water Bodies," ArcGIS model, last modified January 16, 2012. As of July 2, 2012:
http://www.arcgis.com/home/item.html?id=84e780692f644e2d93cefc80ae1eba3a

Finney, Mark A., "A Prototype Simulation System for Large Fire Planning in FPA," U.S. Department of Agriculture, Forest Service, July 5, 2007. As of July 2, 2012:
http://www.fpa.nific.gov/Library/Docs/Science/FPA_SimulationPrototype_0705.pdf

Finney, Mark A., Isaac C. Grenfell, and Charles W. McHugh, "Modeling Containment of Large Wildfires Using Generalized Linear Mixed-Model Analysis," *Forest Science*, Vol. 55, No. 3, June 2009, pp. 249–255.

Fire Program Analysis System, National Interagency Fire Center, *Reference: An Analysis of the Fireline Production Rates Applied to Aerial Retardant Drops Contained in MNIAAPC*, November 1995. As of July 2, 2012:
http://www.fpa.nific.gov/Documents/Library/papers/RetPRates01.pdf

———, "FIG Historical Fire Locations," raw data, December 28, 2009. As of July 2, 2012:
http://www.fpa.nific.gov/Implementation/index_data_downloads.html

———, "Understanding the Fire Program Analysis (FPA) Fire Ignition Generator (FIG)" memorandum, November 9, 2011. As of July 2, 2012:
http://www.fpa.nific.gov/Documents/Library/papers/TP_FIG_20111109.pdf

———, "FPA Model Components and Functionality," undated. As of July 2, 2012:
http://www.fpa.nific.gov/Library/Docs/Science/Web_FPA_Attachments_Components_2006_11_22.pdf

Fire Program Solutions, *Wildland Fire Management Aerial Application Study: Final Report*, Sandy, Oreg., October 17, 2005.

Fried, Jeremy S., and Burton D. Fried, "Simulating Wildfire Containment with Realistic Tactics," *Forest Science*, Vol. 42, No. 3, August 1996, pp. 267–279.

Fried, Jeremy S., J. Keith Gilless, and James Spero, "Analysing Initial Attack on Wildland Fires Using Stochastic Simulation," *International Journal of Wildland Fire*, Vol. 15, No. 1, 2006, pp. 137–146.

Fulé, Peter Z., Amy E. M. Waltz, W. Wallace Covington, and Thomas A. Heinlein, "Measuring Forest Restoration Effectiveness in Reducing Hazardous Fuels," *Journal of Forestry*, Vol. 99, No. 11, November 2001, pp. 24–29.

Gebert, Krista M., economist, *Economic Aspects of Forest Management on Public Lands Unit*, U.S. Department of Agriculture, Forest Service, Rocky Mountain Research Station, personal communication and data, March 2011.

Gebert, Krista M., David E. Calkin, Robert J. Huggett, Jr., and Karen L. Abt, "Economic Analysis of Federal Wildfire Management Programs," in Thomas P. Holmes, Jeffrey P. Prestemon, and Karen L. Abt, eds., *The Economics of Forest Disturbances: Wildfires, Storms, and Invasive Species*, London: Springer, 2008, pp. 295–322.

Gebert, Krista M., David E. Calkin, and Jonathan Yoder, "Estimating Suppression Expenditures for Individual Large Wildland Fires," *Western Journal of Applied Forestry*, Vol. 22, No. 3, July 2007, pp. 188–196.

Gebert, Krista M., and Ervin G. Schuster, "Forest Service Fire Suppression Expenditures in the Southwest," in *Proceedings of the 2002 Fire Conference: Managing Fire and Fuels in the Remaining Wildlands and Open Spaces of the Southwestern United States*, Albany, Calif.: U.S. Department of Agriculture, Forest Service, Pacific Southwest Research Station, General Technical Report PSW-GTR-189, 2008, pp. 227–236.

Gilless, J. Keith, and Jeremy S. Fried, "Stochastic Representation of Fire Behavior in a Wildland Fire Protection Planning Model for California," *Forest Science*, Vol. 45, No. 4, November 1999, pp. 492–499.

Gonsalves, John R., Bombardier Aerospace, "The Bombardier 415—An Historic Perspective," paper presented at the Anaheim Aerial Firefighting Conference and Exhibition, February 19–20, 2009.

Gorte, Ross W., *Forest Fire/Wildfire Protection*, Washington, D.C.: Congressional Research Service, RL30755, January 18, 2006. As of July 2, 2012: http://fpc.state.gov/documents/organization/60721.pdf

Graham, Russell T., ed., *Hayman Fire Case Study*, Ogden, Utah: U.S. Department of Agriculture, Forest Service, Rocky Mountain Research Station, General Technical Report RMRS-GTR-114, 2003. As of July 2, 2012: http://www.fs.fed.us/rm/pubs/rmrs_gtr114.pdf

Gude, Patricia H., J. Anthony Cookson, Mark C. Greenwood, and Mark Haggerty, "Homes in Wildfire-Prone Areas: An Empirical Analysis of Wildfire Suppression Costs and Climate Change," Bozeman, Mont.: Headwaters Economics, 2009. As of July 2, 2012:
http://www.headwaterseconomics.org/wildfire/Gude_Manuscript_4-24-09_Color.pdf

Haight, Robert G., and Jeremy S. Fried, "Deploying Wildland Fire Suppression Resources with a Scenario-Based Standard Response Model," *INFOR*, Vol. 45, No. 1, February 2007, pp. 31–39.

Hammer, Roger B., Susan I. Stewart, and Volker C. Radeloff, "Demographic Trends, the Wildland-Urban Interface, and Wildfire Management," *Society and Natural Resources*, Vol. 22, No. 8, 2009, pp. 777–782.

Hesseln, Hayley, and Douglas B. Rideout, "Using Control Theory to Model the Long-Term Economic Effects of Wildfire," in *Proceedings of the Symposium on Fire Economics, Planning and Policy: Bottom Lines*, Albany, Calif.: U.S. Department of Agriculture, Forest Service, Pacific Southwest Research Station, General Technical Report PSW-GTR-173, 1999, pp. 107–114. As of July 2, 2012:
http://www.fs.fed.us/psw/publications/documents/psw_gtr173/psw_gtr173.pdf

Independent Large Wildfire Cost Panel, *2007 U.S. Forest Service and Department of Interior Large Wildfire Cost Review: Assessing Progress Towards an Integrated Risk and Cost Fire Management Strategy*, April 24, 2008. As of July 2, 2012:
http://www.fs.fed.us/fire/publications/ilwc-panel/report-2007.pdf

Insurance Information Institute, "Wildfire Facts," web page, undated. As of July 2, 2012:
http://www2.iii.org/index.cfm?instanceID=240030

International Code Council, *The Blue Ribbon Panel Report on Wildland Urban Interface Fire*, Washington, D.C., April 2008. As of July 2, 2012:
http://www.iccsafe.org/gr/WUIC/Pages/BlueRibbon.aspx

Kliment, Linda K., Kamran Rokhsaz, John Nelson, and James Newcomb, "Operational Loads on Heavy Air Tankers," extended abstract, 2010.

Large-Cost Fire Independent Review Panel, *Fiscal Year 2008 Large-Cost Fire Independent Review*, Washington, D.C.: U.S. Secretary of Agriculture, 2009.

Liang, Jingjing, Dave E. Calkin, Krista M. Gebert, Tyrone J. Venn, and Robin P. Silverstein, "Factors Influencing Large Wildland Fire Suppression Expenditures," *International Journal of Wildland Fire*, Vol. 17, No. 5, 2008, pp. 650–659.

Linse, Paul, U.S. Forest Service, personal communication, July 17, 2012.

Loane, Ingram Thomas, and James Stanley Gould, *Aerial Suppression of Bushfires: Cost-Benefit Study for Victoria*, Canberra, Australia: National Bushfire Research Unit, Commonwealth Scientific and Industrial Research Organisation, Division of Forest Research, April 1986.

Loomis, John B., and Armando Gonzalez-Caban, "A Willingness-to-Pay Function for Protecting Acres of Spotted Owl Habitat from Fire," *Ecological Economics*, Vol. 25, No. 3, June 1998, pp. 315–322.

Lynch, Dennis L., "What Do Forest Fires Really Cost?" *Journal of Forestry*, Vol. 102, No. 6, September 2004, pp. 42–49.

Martell, D. L., R. J. Drysdale, G. E. Doan, and D. Boychuk, "An Evaluation of Forest Fire Initial Attack Resources," *Interfaces*, Vol. 14, No. 5, September–October 1984, pp. 20–32.

Martinson, Erik J., and Phillip N. Omi, "Performance of Fuel Treatments Subjected to Wildfires," in *Fire, Fuel Treatments, and Ecological Restoration: Conference Proceedings*, Fort Collins, Colo.: U.S. Department of Agriculture, Forest Service, Rocky Mountain Research Station, Proceedings RMRS-P-29, 2003, pp. 7–14. As of July 2, 2012:
http://www.fs.fed.us/rm/pubs/rmrs_p029/rmrs_p029_007_014.pdf

Martinson, Erik, Phillip N. Omi, and Wayne Shepperd, "Effects of Fuel Treatments on Fire Severity," in Russell T. Graham, ed., *Hayman Fire Case Study*, Ogden, Utah: U.S. Department of Agriculture, Forest Service, Rocky Mountain Research Station General Technical Report RMRS-GTR-114, September 2003, pp. 96–126. As of July 2, 2012:
http://www.fs.fed.us/rm/pubs/rmrs_gtr114/rmrs_gtr114_096_126.pdf

McCarthy, Gregory J., *Effectiveness of Aircraft Operations by the Department of Natural Resources and Environment and the Country Fire Authority, 1997–1998*, Beecroft, New South Wales, Australia: Forest Science Centre, Research Report No. 52, July 2003.

Mercer, D. Evan, Jeffrey P. Prestemon, David T. Butry, and John M. Pye, "Evaluating Alternative Prescribed Burning Policies to Reduce Net Economic Damages from Wildfire," *American Journal of Agricultural Economics*, Vol. 89, No. 1, February 2007, pp. 63–77.

Morton, Douglas C., Megan E. Roessing, Ann E. Camp, and Mary L. Tyrrell, *Assessing the Environmental, Social, and Economic Impacts of Wildfire*, New Haven, Conn.: Forest Health Initiative, Global Institute of Sustainable Forestry, Research Paper 001, May 2003. As of July 2, 2012:
http://environment.yale.edu/gisf/files/pdfs/wildfire_report.pdf

Mutch, Robert W., "Fighting Fire with Prescribed Fire: A Return to Ecosystem Health," *Journal of Forestry*, Vol. 92, No. 11, November 1994, pp. 31–33.

National Interagency Aviation Council, *Interagency Aerial Supervision Guide*, NFES 2544, 2009.

National Interagency Coordination Center, National Interagency Fire Center, *Wildland Fire Summary and Statistics Annual Report 2005,* 2005. As of July 2, 2012:
http://www.predictiveservices.nifc.gov/intelligence/2005_statssumm/ 2005Stats&Summ.html

———, *Wildland Fire Summary and Statistics Annual Report 2006,* 2006. As of July 2, 2012:
http://www.predictiveservices.nifc.gov/intelligence/2006_statssumm/ 2006Stats&Summ.html

———, *Wildland Fire Summary and Statistics Annual Report 2007,* 2007. As of July 2, 2012:
http://www.predictiveservices.nifc.gov/intelligence/2007_statssumm/ 2007Stats&Summ.html

———, *Wildland Fire Summary and Statistics Annual Report 2008,* 2008. As of July 2, 2012:
http://www.predictiveservices.nifc.gov/intelligence/2008_statssumm/ 2008Stats&Summ.html

———, *Wildland Fire Summary and Statistics Annual Report 2009,* 2009. As of July 2, 2012:
http://www.predictiveservices.nifc.gov/intelligence/2009_statssumm/ 2009Stats&Summ.html

National Transportation Safety Board, "Safety Recommendation, In Reply Refer to: A-04-29 Through -33," Washington, D.C., April 23, 2004.

National Wildfire Coordinating Group, *Fireline Handbook,* Handbook 3, PMS 410-1, NFES 0065, March 2004. As of July 2, 2012:
http://www.nwcg.gov/pms/pubs/410-1/410-1.pdf

———, *National Wildland Firefighting Fatalities in the United States, 1990–2006,* PMS-841, August 2007. As of July 2, 2012:
http://www.nwcg.gov/pms/pubs/pms841/pms841_all-72dpi.pdf

———, *Quadrennial Fire Review 2009: Final Report,* January 2009.

———, *Glossary of Wildland Fire Terminology,* PMS 205, May 2011. As of July 2, 2012:
http://www.nwcg.gov/pms/pubs/glossary/index.htm

National Interagency Fire Center, data provided to RAND, June 7, 2010.

———, "About NIFC," web page, undated(a). As of July 2, 2012:
http://www.nifc.gov/aboutNIFC/about_main.html

———, "Geographic Area Coordination Centers," homepage, undated(b). As of July 2, 2012:
http://gacc.nifc.gov

NIFC—*See* National Interagency Fire Center.

Office of Management and Budget, "Discount Rates for Cost-Effectiveness, Lease-Purchase, and Related Analyses," Appendix C in Circular No. A-94, Washington, D.C., December 2009.

Plucinski, Matt, "Assessing Aerial Suppression Drop Effectiveness," *Fire Note*, No. 38, September 2009. As of July 2, 2012:
http://www.bushfirecrc.com/managed/resource/0909__firenote38_lowres.pdf

Plucinski, M., J. Gould, G. McCarthy, and J. Hollis, *Effectiveness and Efficiency of Aerial Firefighting in Australia, Part 1*, East Melbourne, Victoria, Australia: Bushfire Cooperative Research Centre, Technical Report Number A0701, June 2007.

Potts, Donald F., David L. Peterson, and Hans R. Zurring, *Watershed Modeling for Fire Management Planning in the Northern Rocky Mountains*, U.S. Department of Agriculture, Forest Service, Pacific Southwest Forest and Range Experiment Station, Research Paper PSW-177, 1985.

Rideout, Douglas B., and Pamela S. Ziesler, "Three Great Myths of Wildland Fire Management," in *Proceedings of the Second International Symposium on Fire Economics, Planning, and Policy: A Global View*, U.S. Department of Agriculture, Forest Service, Pacific Southwest Research Station, General Technical Report PSW-GTR-208, 2008, pp. 319–325.

Rittmaster, R., W. L. Adamowicz, B. Amiro, and R. T. Pelletier, "Economic Analysis of Health Effects from Forest Fires," *Canadian Journal of Forest Research*, Vol. 36, No. 4, 2006, pp. 868–877.

Scott, Joe H., and Robert E. Burgan, *Standard Fire Behavior Fuel Models: A Comprehensive Set for Use with Rothermel's Surface Fire Spread Model*, Fort Collins, Colo.: U.S. Department of Agriculture, Forest Service, Rocky Mountain Research Station, General Technical Report RMRS-GTR-153, June 2005. As of July 2, 2012:
http://www.fs.fed.us/rm/pubs/rmrs_gtr153.pdf

Servis, Steve, and Paul F. Boucher, "Restoring Fire to Southwestern Ecosystems: Is It Worth It?" in *Proceedings of the Symposium on Fire Economics, Planning, and Policy: Bottom Lines*, Albany, Calif.: U.S. Department of Agriculture, Forest Service, Pacific Southwest Research Station, General Technical Report PSW-GTR-173, December 1999, pp. 247–253. As of July 2, 2012:
As of July 2, 2012:
http://www.fs.fed.us/psw/publications/documents/psw_gtr173/psw_gtr173.pdf

Sorenson, B., M. Fuss, Z. Mulla, W. Bigler, S. Wiersma, and R. Hopkins, "Surveillance of Morbidity During Wildfires—Central Florida 1998," *Morbidity and Morality Weekly Report*, Vol. 48, No. 4, 1999, pp. 78–79.

U.S. Air Force, Air Force Total Ownership Cost System, database.

U.S. Department of Agriculture, Forest Service, *National Study of Type I and II Helicopters to Support Large Fire Suppression, Final Report*, 1992.

―――, *Military Airborne Firefighting System 2nd Generation*, December 2004.

―――, *Fiscal Year 2006 President's Budget: Budget Justification*, 2005. As of July 2, 2012:
http://www.fs.fed.us/aboutus/budget

―――, *Fiscal Year 2008 President's Budget: Budget Justification*, 2007. As of July 2, 2012:
http://www.fs.fed.us/aboutus/budget

――――, *Fiscal Year 2010 President's Budget: Budget Justification*, 2009. As of July 2, 2012:
http://www.fs.fed.us/aboutus/budget

―――, *Fiscal Year 2011 President's Budget: Budget Justification*, 2010. As of July 2, 2012:
http://www.fs.fed.us/aboutus/budget/

U.S. Department of Agriculture, Forest Service, Wildland Fire Chemical Systems, "Definitions of Product Types," web page, last modified September 26, 2011. As of July 2, 2012:
http://www.fs.fed.us/rm/fire/wfcs/define.htm

U.S. Department of Agriculture, Forest Service, and U.S. Department of the Interior, *National Study of Airtankers to Support Initial Attack and Large Fire Suppression: Final Report, Phase 1*, March 1995.

―――, *National Study of (Large) Airtankers to Support Initial Attack and Large Fire Suppression: Final Report Phase 2*, November 1996.

U.S. Department of Agriculture, Office of Inspector General, Western Region, *Audit Report: Forest Service Large Fire Suppression Costs*, Report No. 08601-44-SF, November 2006.

―――, *Audit Report: Forest Service's Replacement Plan for Firefighting Aerial Resources*, Report No. 08601-53-SF, July 2009.

U.S. Department of Commerce, Bureau of Economic Analysis, "National Income and Product Accounts, Table 1.1.9, Implicit Price Deflators for Gross Domestic Product," revised March 26, 2010.

U.S. Department of Defense, Defense Acquisition Management Information Retrieval, *Selected Acquisition Report (SAR): C-130J*, December 31, 2007. As of July 2, 2012:
http://www.dod.gov/pubs/foi/logistics_material_readiness/acq_bud_fin/SARs/DEC%202011%20SAR/C-130J%20-%20SAR%20-%2031%20DEC%202011.pdf

Venn, Tyron J., and David E. Calkin, "Challenges of Accommodating Non-Market Values in Evaluation of Wildfire Suppression in the United States," paper presented at the American Agricultural Economic Association annual meeting, Portland, Oreg., July 29–August 1, 2007.

Viscusi, W. Kip, and Joseph E. Aldy, "The Value of a Statistical Life: A Critical Review of Market Estimates Throughout the World," *Journal of Risk and Uncertainty*, Vol. 27, No. 1, August 2003, pp. 5–76.